T0137337

Evolutionary Computation and Complex Networks

Jing Liu · Hussein A. Abbass
Kay Chen Tan

Evolutionary Computation and Complex Networks

 Springer

Jing Liu
Key Laboratory of Intelligent Perception
 and Image Understanding of Ministry
 of Education
Xidian University
Xi'an, Shaanxi, China

Kay Chen Tan
Department of Computer Science
City University of Hong Kong
Kowloon Tong, Kowloon, Hong Kong SAR

Hussein A. Abbass
School of Engineering and Information
 Technology
University of New South Wales
Canberra, ACT, Australia

ISBN 978-3-030-09649-6   ISBN 978-3-319-60000-0   (eBook)
https://doi.org/10.1007/978-3-319-60000-0

This Springer imprint is published by the registered company Springer Nature Switzerland AG
The registered company address is: Gewerbestrasse 11, 6330 Cham, Switzerland

To our beloved families, students and friends

Preface

Evolutionary computation and complex networks have received considerable attention in recent years. Inspired by natural selection and evolutionary theory, evolutionary algorithms (EAs) are optimization heuristics designed to solve optimization problems. They are different from classic optimization algorithms in their reliance on a population of solutions instead of searching using a single solution at a time and their ability to solve nondifferentiable functions with an arbitrary level of complexity. They have been very successful in solving various engineering applications.

Complex networks abstract a wide range of biological and social systems including the Internet, e-mail interactions, gene regulatory networks, science collaboration networks, phone call networks, and citation networks. Because of their significant contributions to our understanding of complex systems, complex networks have been attracting much interest and seen significant advances over the last decade.

Most of the work in the field of complex networks has concentrated on two aspects. On the one hand, research has focused on analyzing properties found in real-world networks, and then designing algorithms to build understanding of the mechanisms underlying the emergence of these properties; that is to say, to design methods that can generate the networks with certain properties. This process is modeled as a form of complex optimization problems. On the other hand, considering the fact that EAs rely on a population of solutions, namely a set of individual solutions, the performance of EAs heavily depends on the interaction among these individuals. Complex network analysis could in principle represent this interaction to understand evolutionary dynamics and how fitter solutions emerge.

The use of complex networks in the field of evolutionary computation have been subjected to a considerable amount of research. The majority of research in this area attempts using complex networks, such as small-world and scale-free networks, as the underlying population architecture in evolutionary algorithms. Moreover, complex networks are also used to analyze fitness landscapes and designing predictive problem difficulty measures. Simultaneously, EAs were also used to solve optimization problems in the field of complex networks, such as community detection problems, network robustness optimization problems, and network reconstruction.

Up to now, the two fields have been treated as two different fields despite their mutual dependency on each other and the similarities sitting underneath their design. This book brings the two fields together to have a single treatment of the topic. The book is structured in three parts. The first part offers a brief introduction of each topic to lay out fundamental background information for the reader. The second part focuses on the application of complex networks to evolutionary computation, that is, summarizing the research studies that have been conducted on how complex networks are used to analyze and improve the performance of EAs. The third part focuses on the complimentary side of how evolutionary computation methods are used to solve optimization problems in complex networks.

In both parts, the authors bring together their own research on the topic to offer a comprehensive treatment under a single roof. We hope that this approach will be useful for researchers and students to offer a unified treatment of the topic in a single source.

Xi'an, China   Jing Liu
Canberra, Australia   Hussein A. Abbass
Kowloon Tong, Hong Kong SAR   Kay Chen Tan
January 2018

Acknowledgements

Writing a book is a journey. Along the way, the authors encounter people and organizations without whom the journey would have been longer and possibly the goal would not have been reached.

We wish to thank our families. The time taken to put this book together could have been well-spent with them. We thank them for allowing us this time to bring this book to life.

The authors wish to thank their organizations for allowing them the time and capacity to complete this book project successfully: Xidian University, University of New South Wales, and the City University of Hong Kong.

This book is based on our own research and research of others. In particular, a special thanks to Prof. David Green from Monash University, Australia, for a collaboration with the first two authors that resulted in their previous book on Dual Phase Evolution, a technique we also used in this book.

While the book summarizes our previous research, that research was funded by a number of organizations including the National Natural Science Foundation of China (NSFC) and the Australian Research Council.

Last but not least, we wish to thank our teachers, students, collaborators, and all those who influenced our knowledge to create the knowledge needed to write this book.

Contents

List of Figures

List of Tables

Part I
Introducing Evolutionary Algorithms and Complex Networks

Chapter 1
Evolutionary Computation

Evolutionary algorithms (EAs) are heuristics designed to solve optimization problems. They are different from classic optimization algorithms in their reliance on a population of solutions instead of searching using a single solution at a time and their ability to solve nondifferentiable functions with arbitrary level of complexity. They have been very successful in solving various engineering applications. The remaining of this chapter offers an introduction to EAs and a number of derivatives of the evolutionary approaches.

1.1 The Basic Evolutionary Algorithm

Early studies of EAs can be traced back to 1960s. These early studies focused on problem domains where a concrete answer is not necessarily known in advance or even well-defined as in the case of designing adaptive systems [8], pattern recognition tasks [11], and functional optimization [35]. In 1975, Holland and DeJong summarized the basic concepts of EAs [21, 34] and laid out the foundations of EAs. In subsequent studies, EAs have been applied into broader engineering fields, including machine learning [27], computational intelligence [32], and parallel systems [75].

Classical EAs have demonstrated successful uses within engineering optimization, especially in problems that are black box, nondifferentiable, multi-objective, and/or with severe uncertainties. Below is a crisp overview of classic optimization.

A simple optimization strategy is to exhaustively search the space of all solutions. Clearly this is an expensive process and does not scale beyond toy problems. Hill climbing is a simple strategy that selects the most promising direction to move to within the search space and continues to do so till no direction exists that can improve the current solution. This form of greedy search can easily get stuck in a local optimum or a valley. While stochastic search techniques, such as simulated annealing and tabu search, rely on moving from a single solution to another one at a

© Springer Nature Switzerland AG 2019
J. Liu et al., *Evolutionary Computation and Complex Networks*,
https://doi.org/10.1007/978-3-319-60000-0_1

time, these techniques avoid the opportunity afforded by an evolutionary population that undertakes implicit parallel search.

Evolutionary algorithms borrow biological concepts and adopt them to engineering science. They have become one of the most important computational intelligence techniques in this century. Based on Darwin's theory, evolutionary algorithms imitate the progress of evolution to find the best solutions in the searching area, and are suitable to resolve complex and nonlinear problems. These problems are classically hard for traditional optimization methods. In 1975, Prof. J. Holland proposed the first model of an evolutionary algorithm [34] and initiated the studies on this widely used computational method.

EAs are stochastic optimization algorithms that rely on a population of solutions and the exchange of information among these solutions to create new ones with an aim to improve solutions over generations. The basic ingredients of EAs are (1) variable encoding, (2) the set of initial population, (3) the design of fitness function, (4) the design of genetic operations, and (5) the configuration of control parameters, including population size, the rate for applying each control parameter, and the possibility of conducting genetic operations.

An EA starts with a sample of solutions called a population. The algorithm then iterates through a number of steps starting with the evaluation of these solutions, selection of solutions to reproduce, mixing of these solutions to allow for information exchange to occur, then the cycle either terminates if termination criteria are satisfied or continues. This process tends to favor the survival of fitter solutions and the elimination of less promising ones.

Information exchange is carried out using what is known as genetic operators. These operators include selection, crossover, and mutation. They contribute to the convergence of the searching process and steer individual solutions toward the optimal solution.

Figure 1.1 illustrates the process of classical EAs. Here, we give detailed explanations of the operators in Fig. 1.1.

- **Encoding**:
  Classical EAs relied on different encoding schemes to transform the variables to be optimized into a representation to operate upon. For example, classic genetic algorithms rely on binary encoding, which offers an opportunity to operate on the level of a bit with logical operators that can speed up processing. However, most implementations did not rely on this characteristic and thus, the encoding was challenged and questioned. Other encoding schemes included the use of real numbers and strings. Some implementations demanded that the length of the encoding string, also known as chromosome, is fixed, while others allowed for variable-length encoding.
- **Initialization**:
  The first initial population gets normally created through random initialization. The rationale behind this approach is to reduce bias in the initial population and create diverse solutions. However, with the finite nature of the population in an EA, initialization bias increases as the population size decreases. Analysts could

Fig. 1.1 The flowchart of
classical EAs

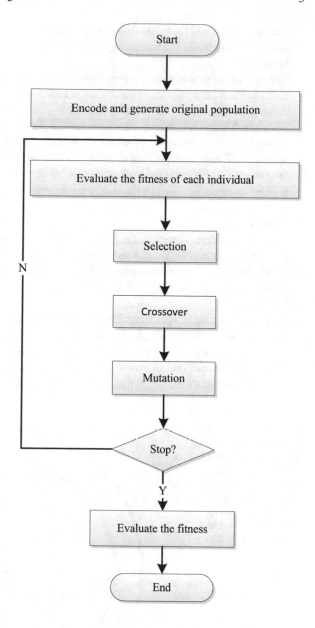

use any initialization approach they see useful for their problems including the use
of domain knowledge to seed the population with specific solutions.

- **Fitness Evaluation**:
  Fitness is classically known as the performance of an individual relative to the
  population. Optimization problems would include one or more objective functions.
  These objective functions are used to measure the performance of an individual

solution. Fitness is then calculated in different manners to estimate the quality of a solution relative to the rest of the population.

- **Selection**:
The purpose of selection is to choose promising individuals, which are more likely to contribute to future improvements in the fitness of the children they produce. The concept of selection is inspired by Darwin's "survival of the fittest" evolutionary principle. Computationally, selection gets implemented either deterministically or stochastically. An example of a simple method for selection is *roulette wheel selection*, where we should first calculate the sum of all objective values in the population ($\sum f$). Each individual is selected in proportion to the fraction it contributes to this sum ($\frac{f_i}{\sum f}$). In Fig. 1.2, a simple example of *roulette wheel selection* is given.

- **Crossover**:
The concept of crossover can simply be seen as a process for information exchange. Normally it requires a pair of individuals. Children are then made by combining information from the pair. In classic EAs, crossover is a key operator and plays a significant role in the design of the algorithm.

- **Mutation**:
Mutation introduces variations into the population of solutions by injecting new information into each individual. The quality of these new information is left to the selection operator to act on. However, the key idea is to ensure that new sources of information get introduced to avoid getting stuck in a local optimum. Taking binary coding as an example, the mutation operation is conducted on the bits with

Fig. 1.2 A simple example of *roulette wheel selection*. There are four individuals in the population, and the fraction of each individual is depicted in the figure. In the selection process, the possibility of choosing is based on the fraction

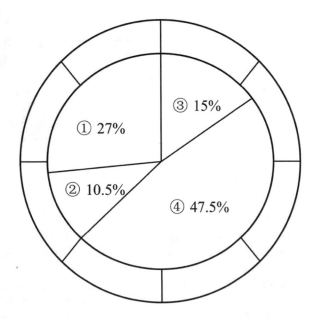

a relatively small possibility (i.e., randomly choose a bit of an individual), then flip the value of this bit to generate a new individual.

The above operations capture the basic ingredients of the classical EA. A simple implementation for a single-objective unconstrained optimization problem would use the following setup: randomly initialize the population; evaluate individual solutions using the objective function; use *roulette wheel selection* to select pairs of parents; crossover each pair of parents to create a child; mutate each child; evaluate the children and repeat the cycle.

Many variations of the classical EAs have emerged over the years. Some are grounded in evolutionary theory, while others are grounded in physics or computational sciences. We introduce examples of these. Differential evolution relies on concepts of selection, but the crossover and mutation operators do not resemble any biological process. Instead, they could be seen as approximate gradient information. Memetic algorithms draw inspirations from Dawkins's selfish gene and his use of the concept of a "meme" to represent the unit of cultural inheritance similar to a "gene" representing the unit of biological inheritance. Particle swarm optimization draws inspiration from physics and represents each solution as a particle with a velocity vector and a model of the dynamics of the particles. These three approaches are today the most popular approaches to solve engineering optimization problems.

1.2 Differential Evolution

Differential evolution (DE) was originally proposed by Storn and Price in 1995 [77]. Soon after that, the strength of DE was demonstrated at the first and the second International Conference on Evolutionary Computation [62, 78]. This strength stems from the fact that classic DE does not rely on sorting or any computationally expensive operations; thus the algorithm is fast and when parallelized, it becomes even superfast. Its reliance on vectors makes it ideal for graphical processing units (GPUs) implementations, making it superfast. Moreover, by using floating point representations, it is space-efficient when it comes to representing most parameter optimization problems. In practice, DE has obtained high-quality solutions for practical problems, making it an ideal optimization heuristic for most engineering problems.

The huge success of DE has attracted attentions of researchers from various domains. Many DE variants have been proposed to make the algorithm more and more powerful [6, 18, 23, 37, 54, 63–65, 68, 79, 83].

In this section, some basic concepts of the classical DE are introduced along with some important DE variants that were introduced in recent years and some potential future research directions.

1.2.1 Basic Concepts

DE is designed to deal with the classical problem of search and optimization. These problems can be modeled as finding a vector $X = [x_1, x_2, \ldots, x_D]^T$ which minimizes or maximizes an objective or fitness function[1] $f(X) : \Omega \subseteq \Re^D \to \Re$. Ω is the problem's domain and spans $\Re^D$ for unconstrained optimization problems. Since $\max\{f(X)\} = \min\{-f(X)\}$, the canonical form uses a minimization problem without loss of generality. DE searches for global optimum in a D-dimensional real number space $\Re^D$. It is an iterative algorithm presented in Fig. 1.4. Different strategies used in each step of DE create different DE variants. Below, we describe the original DE proposed by Storn and Price [77].

Initialization

DE uses NP vectors of D-dimensional real-valued parameters as a population in each generation G, which is defined in Eq. 2.1.

$$X_{i,G} = [x_{1,i,G}, x_{2,i,G}, \ldots, x_{D,i,G}] \quad i = 1, 2, \ldots, NP \qquad (1.1)$$

The initialization stage creates the initial population ($G = 0$). Given the predefined lower and upper bounds of the search space: $X_{\min} = \{x_{1,\min}, x_{2,\min}, \ldots, x_{D,\min}\}$ and $X_{\max} = \{x_{1,\max}, x_{2,\max}, \ldots, x_{D,\max}\}$, the initial population can be generated as

$$X_{j,i,0} = x_{j,\min} + \text{rand}_{ij}() \cdot (x_{j,\max} - x_{j,\min}) \qquad (1.2)$$

where $\text{rand}_{ij}()$ generates a uniformly distributed random number between 0 and 1. This initialization randomizes individuals to uniformly cover the search space as much as possible. Assuming unbiased random number generator, the larger the population, the more likely it uniformly covers the search space.

Mutation

For vector $X_{i,G}$, a mutant vector $V_{i,G}$ is generated according to

$$V_{i,G} = X_{r_1,G} + F \cdot (X_{r_2,G} - X_{r_3,G}) \qquad (1.3)$$

where r_1, r_2, and r_3 are random indexes chosen without replacement from the range $[1, NP]$. F is a constant factor in the range of $[0, 2]$ which represents the step length in classical optimization responsible for the amplification of the differential variation. The process is illustrated in Fig. 1.3 as a 2-D example.

---

[1] We will not differentiate between objective and fitness functions in parameter optimization problems in this book.

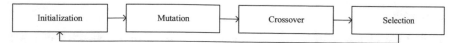

Fig. 1.3 A simple DE mutation in 2-*D* space

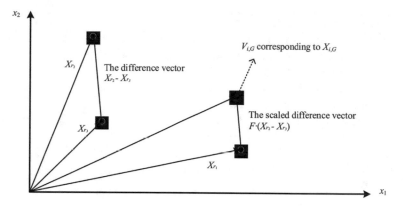

Fig. 1.4 Main stage of DE algorithm

Crossover

Crossover is introduced to increase the diversity of the population. The scheme of generating a target vector $U_{i,G} = [u_{i,1,G}, u_{i,2,G}, \ldots, u_{i,D,G}]$ can be outlined as

$$u_{i,j,G} = \begin{cases} v_{j,i,G} & \text{if } \mathrm{rand}_{ij}() \leq Cr \text{ or } j = j_{\mathrm{rand}} \\ x_{j,i,G} & \text{otherwise} \end{cases} \tag{1.4}$$

where $\mathrm{rand}_{ij}()$ generates a uniformly distributed random number between 0 and 1. Cr is the crossover constant, which indicates the probability of the event that the component of U will be inherited from V. j_{rand} is a randomly chosen index from $[1, 2, \ldots, D]$, which ensures $U_{i,G}$ gets at least one component from $V_{i,G}$.

Selection

To keep the size of the population constant in each generation, the selection operation determines whether the newly generated vector $U_{i,G}$ or the original vector $X_{i,G}$ survives to the next generation. This operation is described as

$$X_{i,G+1} = \begin{cases} U_{i,G} & \text{if } f(U_{i,G}) \leq f(X_{i,G}) \\ X_{i,G} & \text{otherwise} \end{cases} \tag{1.5}$$

where $f()$ is the objective function to be minimized. The newly generated vector survives only if it yields a better value of the objective function; otherwise, the original vector is retained.

Summary

An iteration of the classical DE consists of the steps of mutation, crossover, and selection. The pseudo-code for DE is shown in Algorithm 1.1.

Algorithm 1.1 Pseudo-code for DE

Set the parameters of DE: population size NP, scale factor F, crossover rate Cr.
Set current generation $G = 0$ and initialize a generation as Eq. 1.2.
while the stopping criterion is not satisfied **do**
  for each individual with index i in the population **do**
    Generate $V_{i,G}$ according to Eq. 1.3
    Generate $U_{i,G}$ according to Eq. 1.4
    Generate $X_{i,G}$ according to Eq. 1.5
  end for
  Increase the generation count: $G = G + 1$
end while

1.2.2 Important Variants

Since the proposal of DE in 1995, researchers from different domains have been attracted to this powerful algorithm and created a plenty of variants of the basic DE algorithm. Here, we offer some brief introductions of some important DE variants.

Fan and Lampinen [23] proposed a trigonometric mutation operator to speed up its performance. It uses the scheme in Eq. 1.6 with a probability Γ.

$$
\begin{aligned}
V_{i,G} = {} & (X_{r_1} + X_{r_2} + X_{r_3})/3 + (p_2 - p_1)(X_{r_1} - X_{r_2}) \\
& + (p_3 - p_2)(X_{r_2} - X_{r_3}) + (p_1 - p_3)(X_{r_3} - X_{r_1}) \\
p_i = {} & |f(X_{r_i})| / \sum_{i=1}^{3} |f(X_{r_i})|
\end{aligned}
\tag{1.6}
$$

Price proposed a new crossover strategy named "DE/current-to-rand/1" [63] to make the crossover rotationally invariant. The new linear recombination operator is as follows:

$$
U_{i,G} = X_{i,G} + k_i \cdot (V_{i,G} - X_{i,G})
\tag{1.7}
$$

where k_i is the combination coefficient. The experiments in [63, 64] showed that it is effective to be a uniform random distribution in the range of [0, 1].

Das et al. [18] proposed a DE variant named as DEGL. In his algorithm, he introduced a neighborhood topological structure into DE. DEGL uses the best vector in
neighborhood to approach a mutation operator. By limiting the information spreading
between neighbors, DEGL reduces the chance of sinking in local minima.

Qin et al. [65] proposed a self-adaptive variant of DE (SaDE) to overcome the
issue that some vector generation strategies may be ineffective over certain problems.
In SaDE, four effective mutation strategies constitute a candidate pool. Each strategy
is chosen in mutation with a probability, which is learned from its success rate in
improving solutions within a certain number of the beginning generations.

Abbass et al. [2] proposed the first Pareto-based DE algorithm (PDE) for multi-
objective optimization. Selection is carried out using the concept of Pareto dominance. Its self-adaptive version (SPDE) [1] used a Gaussian distribution to sample
the amplification factor.

1.2.3   Potential Future Research Directions

The research of DE has reached a mature state. However, there are still many open
problems. A survey by Das and Suganthan [19] gave eight potential future research
directions and we briefly summarize their points below.

1. Mutation schemes in DE is additive. Is it possible to design a rotation-based
   mutation operation to improve the explorative power of DE? It is important to
   emphasize that additive operators are more time efficient than multiplicative ones;
   hence, this question needs to be addressed carefully in light of the increase in
   computational cost that may accompany the use of a multiplicative operator.
2. The performance of DE depends on the selected mutation strategy and parameters.
   However, a mutation strategy may fit some specific kinds of problems. If we can
   borrow the ensemble concept from machine learning, we may form an ensemble
   DE approach to solve different problem scenarios.
3. DEGL [18] uses the topological neighborhood of the individuals for devising a
   local mutation scheme. The effect of various neighborhood topologies should be
   investigated and integrated in the future.
4. There is a shortage in developing theoretical underpinnings of the DE family of
   algorithms. Analyzing the probabilistic convergence properties of DE remains
   an open problem even for some basic objective functions.
5. DE's weak selective pressure may lead to an inefficient exploitation especially in
   some rotated optimization problems. Sutton et al. [80] pointed out this drawback
   and proposed a rank-based selection. Other selection scheme like tournament
   selection is worthy to be investigated.
6. In an EA, the fitness proportionate selection without elitism helps the population to escape local optima. However, DE usually uses a very hard selection. For
   this reason, DE is very sensitive to the initial population and may suffer from

premature convergence [76]. It is a very interesting future research topic to combine the advantages of DE with some nonelitist selection in order to reduce the
risk of premature convergence.

7. It is unclear if DE could be as successful on combinatorial optimization problems as it has been on continuous parameter optimization problems. Designing
   operators suitable for discrete optimization is an open area of research within
   DE.

8. Some multi-objective optimization DE variants have been proposed [2, 6, 68].
   However, they may perform poorly over some many-objective optimization problems, which typically deal with more than three objective functions. Extending
   the multi-objective DE variants to many-objective scenarios is another large field
   for future research.

1.3   Memetic Algorithm

It was in late 1980s that the term "*Memetic Algorithms*" (MAs) [52] was used signi-
fying the birth of a new family of meta-heuristics. MAs have several characteristics
that make them effective in dealing with some problems. The adjective "memetic"
comes from the term "meme," coined by Dawkins [67] to denote an analogous to the
gene in the context of cultural evolution.

It is well established that depending on the property and complexity of a problem,
computational methodologies that may have demonstrated performance advantage
on a particular class of problems, they need to trade off their performance degradation on other classes of problems: an observation that has been classically described
as the "no free lunch theorem" [84]. The theorem has served as a key driving force
of computational intelligence research for more than a decade where a plethora of
dedicated MAs have been manually crafted by analysts to solve domain-specific complex problems. The incorporation or embedding of knowledge about the underlying
problem within search algorithms is now generally accepted as being beneficial for
improving search performance [49, 56, 57]. It is worth noting that while it is possible
to use existing search methodologies for general black-box optimization, it comes
with the disadvantage of performing less competitively compared to problem-specific
approaches. A more detailed description of MA is given in Fig. 1.5.

The first step in any population-based heuristic is to initialize the population of
solutions. Problem knowledge can be introduced at this stage by using constructive
heuristics, and operators are designed considering the feature of the problem. While
it is possible to consider local improvement as one of these operators, it plays such
a distinctive role in most MAs that it is independently depicted in the pseudo-code.

Recombination is the algorithmic component that best captures exchange of information among two or more agents in MAs. By using this operation, the relevant information contained in the parents is combined to produce new solutions. Relevance
here amounts to be significant when it comes to evaluating the quality of solutions.

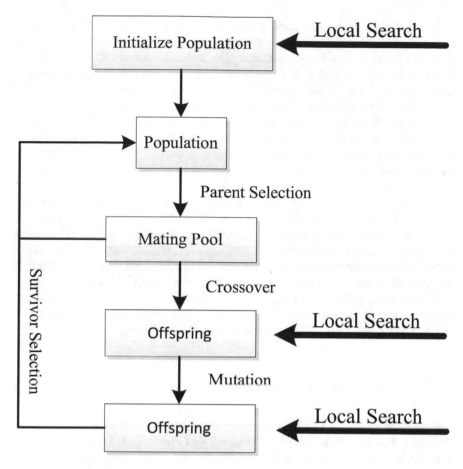

Fig. 1.5 MA procedure

Mutation is another classical reproductive operator. Its role is that of injecting new information in the population (at a lower rate than other operators that exchange information to prevent the search from degradations caused by the stochastic nature of search operators).

The local search algorithm is another operator. For example, it can be used after the utilization of any other recombination or mutation operators; alternatively, it could be just used at the end of the reproductive stage. The utilization of this local-improver is a key characteristic of MAs. It is precisely because of the use of this mechanism for improving individuals on a local (and even autonomous) basis that the term *agent* is used. Thus, the MA can be viewed as a collection of agents performing an autonomous exploration of the search space, cooperating via recombination, and competing for computational resources due to the use of selection/replacement mechanisms.

Within the computational intelligence community, research on MA has since grown significantly and the term has come to be associated with the pairing of meta-heuristics or population-based methodologies with a separate local search process that materializes in various forms of individual learning and social learning. To date, almost all successful stochastic optimization algorithms including meta-heuristics and EAs involve some forms of local search procedure or a meme-learning mechanism in their design. Such synergy has been commonly referred to in the literature as hybrid EAs, such as Baldwinian EAs, Lamarckian EAs, and genetic local searches. Here, the basic unit of cultural transmission, i.e., meme, is regarded by the computational intelligence community as a local search procedure capable of generating refinements on given individual(s).

The use of hybrids algorithms that combines local search with population-based search has ignited the initial interests and studies of memetic computation research. Remarkable success on significant instantiations of specialized MAs [38] across a wide range of application domains has been reported, ranging from NP-hard combinatorial problems such as quadratic assignment [81], permutation flow shop scheduling [33], VLSI floor planning [30], gene/feature selection [91], traveling salesman [47], scheduling and routing [48], multidimensional knapsack [43], data-mining, machine learning and artificial neural networks [5], dynamic optimization problems [10, 46], and computationally expensive environments [58, 59].

1.4   Particle Swarm Optimization

Particle swarm optimization (PSO) is our last global optimization heuristic to be covered in this introductory chapter. Kennedy and Eberhart [41] first proposed the heuristic in 1995 inspired by the simulation of bird flocking. Their work inspired a lot of new intelligent approaches and contributed to the new field known as swarm intelligence (SI).

PSO, similar to other SI methods, is inspired by research in decentralized and self-organized systems. These systems consist of a population of agents which follow some simple common rules. The behaviors of agents are local, random at a certain degree, and without a centralized control structure. However, the consequence of interactions between those agents shows "intelligence" in global scope which is unknown to the individual agents.

Besides PSO, many other SI methods have been proposed. Ant colony optimization (ACO) by Dorigo [22] is inspired by ant colonies. Bacterial foraging optimization (BFO) by Müller et al. [55] and Passion [60] is based on the group foraging behavior of bacteria. Marriage in honey-bees optimization [3, 4] simulates the mating process of honey bees. Krishnanand and Ghose [44] proposed glowworm swarm optimization (GSO) inspired by glowworm carrying luminescence. Bat algorithm (BA) by Yang [86] is inspired by echolocation of microbats.

A survey in [90] gives a figure of the distribution of publications in total and per year of SI algorithms. It is reasonable to say that PSO is the most prevalent SI method. Here in this section, we focus on the introduction of PSO along with its variants and development.

1.4.1 Approach of PSO

PSO searches the solution space via a swarm of particles. These particles update their positions from one iteration to another. In each iteration, a particle moves toward its previously known best position (*pbest*) and the global best position (*gbest*) in the population [89]. The *pbest* and *gbest* are defined as follows.

$$
\text{pbest}(i, t) = \underset{k=1,2,\ldots,t}{\arg\min} f(P_i(k)), \quad i \in [1, 2, \ldots, N_P]
$$
$$
\text{gbest}(i, t) = \underset{\substack{k=1,2,\ldots,t \\ i=1,2,\ldots,N_P}}{\arg\min} f(P_i(k)) \tag{1.8}
$$

where i is the index of the particle in the population of N_P particles, t is the number of the current iteration, f is the objective function to be minimized, and $P_i(k)$ is the position of the ith particle in iteration k.

To update the position of particles, we use the following equations.

$$
V_i(t + 1) = \omega V_i(t) + c_1 r_1 (\text{pbest}(i, t) - P_i(t))
$$
$$
+ c_2 r_2 (\text{gbest}(i, t) - P_i(t)) \tag{1.9}
$$

$$
P_i(t + 1) = P_i(t) + V_i(t + 1) \tag{1.10}
$$

where V denotes the velocity of a particle, ω is an inertia weight factor to balance local exploitation with global exploration, r_1 and r_2 are two uniformly distributed random numbers in the range [0, 1], and c_1 and c_2 are nonnegative control parameters named acceleration coefficients. The pseudo-code of PSO is given in Algorithm 1.2.

1.4.2 Studies on PSO

A major part of the studies of PSO is to modify the method to improve performance. Yang et al. [85] proposed the quantum PSO (QPSO) heuristic, which introduced concepts in quantum mechanics into PSO. Kennedy [40] proposed the bare-bones PSO (BBPSO). BBPSO uses a sampling of a parametric probability density function to update the position and velocity of particles. Chuang et al. [13] integrated the chaotic map in chaos theory with PSO and proposed an improved method named as

---

Algorithm 1.2 Pseudo-code of PSO.

Input: Objective function $f(X) : \mathfrak{R}^N \to \mathfrak{R}$ to minimize.
  The search space $D_x = \{X \,|\, l_1 \leq x_i \leq u_i\}$.
Output: The global optimum X^*.
  for each particle $i = 1, 2, \dots, N_P$ **do**
    Initialize P_i with a uniformly distributed vector in D_x;
    pbest$_i \leftarrow P_i$;
    Initialize V_i with a uniformly distributed vector in $D_V = \{V \,|\, l_i - u_i \leq v_i \leq u_i - l_i\}$;
  end for
  gbest $\leftarrow \text{argmin} f(P_i)$;
  $t \leftarrow 1$;
  while the termination criteria is not satisfied **do**
    for each particle $i = 1, 2, \dots, N_P$ **do**
      Pick two uniformly distributed random number in $[0, 1]$ as r_1 and r_2;
      $V_i \leftarrow \omega V_i + c_1 r_1(\text{pbest}_i - P_i) + c_2 r_2(\text{gbest} - P_i)$;
      $P_i \leftarrow P_i + V_i$;
      if $f(P_i) < f(\text{pbest}_i)$ **then**
        pbest$_i \leftarrow P_i$;
        if $f(P_i) < f(\text{gbest})$ **then**
          gbest $\leftarrow P_i$;
        end if
      end if
    end for
    $t \leftarrow t + 1$;
  end while
  return *gbest* as X^*.

---

chaotic PSO (CPSO). Shi and Eberhart [73] combined fuzzy systems with PSO to make a powerful PSO method named as FPSO.

The classical PSO is designed for single-objective nonconstrained optimization within a continuous space. To extend the application of PSO, a lot of studies have been proposed. Qiu et al. [66] proposed a multiple-objective PSO (MOPSO) method. Daneshyari and Yen [17] proposed a cultural-based constrained PSO to solve constrained optimization problems. Chen and Ludwig [12] proposed an implementation of discrete PSO to solve discrete optimization and integer programming problems.

Another important area of studies on PSO is theoretical analysis of the heuristic itself. Zhang et al. [88] proposed a way to estimate the control parameters using control theory. Kumar and Chaturvedi [45] used a fuzzy-logic controller to adaptively adjust the inertia weight. Kan and Jihong [39] offered a general mathematical description and proved the existence and uniqueness of the convergence position.

1.4.3 Drawbacks and Future Direction

As a SI algorithm, PSO has some drawbacks including premature, and slow convergence, and sensitivity to initial parameter settings. PSO does not have a crossover operator to exchange information between particles. The trade-off between exploitation and exploration is harder to handle.

Future work on PSO could be spent on efforts to understand the reason behind these drawbacks and overcome these difficulties to improve PSO's performance. A systematical study of heterogeneity in PSO is also needed. The parameter selection strategy and the topology used for information exchange are additional dimensions that are open for future investigations.

1.5 Multi-objective Evolutionary Algorithm

EAs have been traditionally used for solving problems with a single objective. However, most real-world problems have multiple conflicting objectives with a set of trade-off solutions. A multi-objective problem (MOP) can be formulated as

$$\min_{w} F(w) = (f_1(w), f_2(w), \dots, f_m(w))^T \tag{1.11}$$

which subjects to $w = (w_1, w_2, \dots, w_n) \in \Gamma$, where w is called the decision vector, and Γ is the feasible region in the decision space. In general, the objectives in a MOP conflict with one another, where a single solution cannot be found in the feasible space that minimizes all the objectives simultaneously. Thus, the concept of Pareto optimal solutions is widely used.

Without loss of generality, we consider a minimization problem. Given two points $w_a, w_b \in \Gamma$, w_a dominates w_b (written as $w_a \succ w_b$), iff $f_i(w_a) \leq f_i(w_b)$ for all $i = 1, 2, \dots, m$, and $f_j(w_a) < f_j(w_b)$ for at least one $j = 1, 2, \dots, m$. The set of all Pareto optimal solutions is called Pareto optimal set which gets defined as follows:

$$PS = \{w \in \Gamma | \neg \exists w \in \Gamma, w^* \prec w\} \tag{1.12}$$

where w^* is a Pareto optimal solution to Eq. 1.11 if there does not exist another solution w in Γ that dominates w^*. The PS in the objective space forms a Pareto frontier (PF) which is defined as

$$PF = \{F(w) | w \in PS\} \tag{1.13}$$

The classical literature on solving multi-objective optimization problems (MOPs) rely on tricks or indirect methods by iterating over a classic single-objective optimizer. This is due to the fact that classical methods normally work with a single solution at a time, whereas EAs work with a sample of solutions concurrently.

During the 70s and 80s, EA researchers realized the need for solving MOPs in practice and mainly resorted to the use of weighted-sum approaches to convert multiple objectives into a single-objective problem [24, 69]. However, it is not obvious how to exactly choose these key parameters so as to trade off performance without heavily relying on a decision maker. The first implementation of a real multi-objective evolutionary algorithm, a vector-evaluated GA (VEGA), was suggested by Schaffer in 1984 [72]. However, VEGA generated solutions but lacked *Diversity* (a set of solutions which is diverse enough to represent the entire range of the Pareto optimal front). Ironically, no significant study was performed for almost a decade, until a revolutionary 10-line sketch of a new nondominated sorting procedure suggested by Goldberg [26].

Goldberg suggested using the concept of *dominance* to assign more copies to nondominated individuals in a population. Since diversity is another concern, he also suggested the use of a *niching* strategy [28] among solutions of a nondominated class. Then, some independent groups of researchers developed different versions of the multi-objective evolutionary algorithms during 1993–1994 [25, 36, 74]. Independently, Poloni [61] suggested a dominance-based multi-objective evolutionary algorithm, called a multi-objective genetic algorithm (MOGA), in which he replaced niching with a toroidal grid-based local selection method to find multiple trade-off solutions. These early multi-objective EAs established the foundations for research in MOEA, but suffered from the fact that they did not use an elite preservation mechanism in their procedures.

Inclusion of elitism in an MOEA provides a monotonically nondegrading performance [70]. The second generation of MOEA implemented an elite-preserving operator in different ways and gave birth to elitism-based MOEA procedures, such as nondominated sorting GA NSGA-II [20], strength Pareto EA (SPEA) [93], and Pareto-archived ES (PAES) [42].

The research continued with a plethora of competent MOEAs, such as the MOEA based on decomposition (MOEA/D) [87]; the strength Pareto evolutionary algorithm (SPEA) and its improved version SPEA2 [92]; the Pareto-archived evolutionary strategy (PAES) and its improved versions Pareto-envelope-based selection algorithm (PESA) and PESA2 [16]; multi-objective messy GA (MOMGA) [82]; multi-objective microGA [14]; neighborhood constrained GA [50]; adaptive range MOGA (ARMOGA) [71]; and others. Moreover, there exist other EA-based methodologies, such as particle swarm-based MOEA [15, 53], ant-based MOEA [29, 51], and Pareto-based Differential Evolution (PDE) [7]. Other heuristics including the simulated annealing method were used to find multiple Pareto optimal solutions for multi-objective optimization problems [9] and the Tabu search method [31].

References

1. Abbass, H.: The self-adaptive pareto differential evolution algorithm. In: Proceedings of the IEEE Congress on Evolutionary Computation (CEC2002), vol. 1, pp. 831–836. IEEE Press, Piscataway, NJ (2002)
2. Abbass, H., Sarker, R., Newton, C.: PDE: A pareto-frontier differential evolution approach for multi-objective optimization problems. In: Proceedings of the IEEE Congress on Evolutionary Computation (CEC2001), vol. 2, pp. 971–978. IEEE Press, Piscataway, NJ (2001)
3. Abbass, H.A.: Mbo: marriage in honey bees optimization-a haplometrosis polygynous swarming approach. In: Proceedings of the 2001 Congress on Evolutionary Computation, vol. 1, pp. 207–214. IEEE (2001)
4. Abbass, H.A.: An agent based approach to 3-SAT using marriage in honey-bees optimization. Int. J. Know. Based Intell. Eng. Syst. **6**(2), 64–71 (2002)
5. Abbass, H.A.: An evolutionary artificial neural networks approach for breast cancer diagnosis. Artif. Intell. Med. **25**(3), 265–281 (2002)
6. Abbass, H.A., Sarker, R.: The pareto differential evolution algorithm. Int. J. Artif. Intell. Tools **11**(04), 531–552 (2002)
7. Abbass, H.A., Sarker, R., Newton, C.: PDE: a pareto frontier differential evolution approach for multi-objective optimization problems. In: Proceedings of the Congress on Evolutionary Computation, vol. 2, pp. 971–978. IEEE Service Center, Seoul Korea (2001)
8. Bagley, J.D.: The behavior of adaptive system which employ genetic and correlation algorithm. Ph.D. thesis, University of Michigan (1967)
9. Bandyopadhyay, S., Saha, S., Maulik, U., Deb, K.: A simulated annealing-based multi-objective optimization algorithm: AMOSA. IEEE Trans. Evol. Comput. **12**(3), 269–283 (2008)
10. Caponio, A., Cascella, G.L., Neri, F., Salvatore, N., Sumner, M.: A fast adaptive memetic algorithm for online and offline control design of PMSM drives. IEEE Trans. Syst. Man Cybern. Part B (Cybern.) **37**(1), 28–41 (2007)
11. Cavicchio, D.J.: Adaptive search using simulated evolution. Ph.D. thesis, University of Michigan (1970)
12. Chen, M., Ludwig, S.A.: Discrete particle swarm optimization with local search strategy for rule classification. In: 2012 Fourth World Congress on Nature and Biologically Inspired Computing (NaBIC), pp. 162–167. IEEE (2012)
13. Chuang, L.Y., Tsai, S.W., Yang, C.H.: Chaotic catfish particle swarm optimization for solving global numerical optimization problems. Appl. Math. Comput. **217**(16), 6900–6916 (2011)
14. Coello, C.A.C., Pulido, G.T., et al.: A micro-genetic algorithm for multi-objective optimization. In: EMO, vol. 1, pp. 126–140. Springer (2001)
15. Coello Coello, C.A.: Mopso: a proposal for multiple objective particle swarm optimization. In: Proceedings of the 2002 Congress on Evolutionary Computation (CEC 2002), vol. 2, pp. 1051–1056 (2002)
16. Corne, D.W., Knowles, J.D., Oates, M.J.: The pareto envelope-based selection algorithm for multi-objective optimization. In: International Conference on Parallel Problem Solving from Nature, pp. 839–848. Springer (2000)
17. Daneshyari, M., Yen, G.G.: Constrained multiple-swarm particle swarm optimization within a cultural framework. IEEE Trans. Syst. Man Cybern. Part A Syst. Hum. **42**(2), 475–490 (2012)
18. Das, S., Abraham, A., Chakraborty, U.K., Konar, A.: Differential evolution using a neighborhood-based mutation operator. IEEE Trans. Evol. Comput. **13**(3), 526–553 (2009)
19. Das, S., Suganthan, P.N.: Differential evolution: a survey of the state-of-the-art. IEEE Trans. Evol. Comput. **15**(1), 4–31 (2011)
20. Deb, K., Pratap, A., Agarwal, S., Meyarivan, T.: A fast and elitist multi-objective genetic algorithm: NSGA-II. IEEE Trans. Evol. Comput. **6**(2), 182–197 (2002)
21. DeJong, K.A.: An analysis of the behavior of a class of genetic adaptive systems. Ph.D. thesis (1975)
22. Dorigo, M.: Optimization, learning and natural algorithms. Ph.D. thesis, Politecnico di Milano, Italy (1992)

23. Fan, H.Y., Lampinen, J.: A trigonometric mutation operation to differential evolution. J. Glob. Optim. **27**(1), 105–129 (2003)
24. Fogel, L.J., Owens, A.J., Walsh, M.J.: Artificial intelligence through simulated evolution (1966)
25. Fonseca, C.M., Fleming, P.J., et al.: Genetic algorithms for multi-objective optimization: formulation discussion and generalization. In: Icga, vol. 93, pp. 416–423 (1993)
26. Goldberg, D.: Genetic algorithms in search, optimization, and machine learning (1989)
27. Goldberg, D.E.: Genetic algorithms in search, optimization, and machine learning. Addison-Wesley, Reading (1989)
28. Goldberg, D.E., Richardson, J., et al.: Genetic algorithms with sharing for multimodal function optimization. In: Genetic Algorithms and their Applications: Proceedings of the Second International Conference on Genetic Algorithms, pp. 41–49. Lawrence Erlbaum, Hillsdale, NJ (1987)
29. Gravel, M., Price, W.L., Gagné, C.: Scheduling continuous casting of aluminum using a multiple objective ant colony optimization metaheuristic. Eur. J. Oper. Res. **143**(1), 218–229 (2002)
30. Gwee, B.H., Lim, M.H.: A GA with heuristic-based decoder for ic floorplanning. Integr. VLSI J. **28**(2), 157–172 (1999)
31. Hansen, M.P.: Tabu search for multi-objective optimization: MOTS. In: Proceedings of the 13th International Conference on Multiple Criteria Decision Making, pp. 574–586 (1997)
32. Harp, S.: Towards the genetic synthesis of neural networks. In: ICGA, pp. 360–369 (1989)
33. Hasan, S.K., Sarker, R., Essam, D., Cornforth, D.: Memetic algorithms for solving job-shop scheduling problems. Memet. Comput. **1**(1), 69–83 (2009)
34. Holland, J.H.: Adaptation in Natural and Artificial Systems: an Introductory Analysis with Application to Biology, Control, and Artificial Intelligence. University of Michigan Press, Ann Arbor, MI (1975)
35. Hollstien, R.B.: Artificial genetic adaptation in computer control systems. Ph.D. thesis, University of Michigan (1971)
36. Horn, J., Nafpliotis, N., Goldberg, D.E.: A niched pareto genetic algorithm for multi-objective optimization. In: Proceedings of the First IEEE Conference on Evolutionary Computation, 1994. IEEE World Congress on Computational Intelligence, pp. 82–87. Ieee (1994)
37. Iorio, A.W., Li, X.: Solving rotated multi-objective optimization problems using differential evolution. In: Australasian Joint Conference on Artificial Intelligence, pp. 861–872. Springer (2004)
38. Jeong, S., Hasegawa, S., Shimoyama, K., Obayashi, S.: Development and investigation of efficient GA/PSO-hybrid algorithm applicable to real-world design optimization. IEEE Comput. Intell. Mag. **4**(3) (2009)
39. Kan, W., Jihong, S.: The convergence basis of particle swarm optimization. In: 2012 International Conference on Industrial Control and Electronics Engineering (ICICEE), pp. 63–66. IEEE (2012)
40. Kennedy, J.: Bare bones particle swarms. In: Proceedings of the 2003 IEEE Swarm Intelligence Symposium. SIS 2003, pp. 80–87. IEEE (2003)
41. Kennedy, J., Eberhart, R.: Particle swarm optimization. In: Proceedings of the IEEE International Conference on Neural Networks, 1995, vol. 4, pp. 1942–1948. IEEE (1995)
42. Knowles, J.D., Corne, D.W.: Approximating the nondominated front using the pareto archived evolution strategy. Evol. Comput. **8**(2), 149–172 (2000)
43. Knowles, J.D., Corne, D.W.: M-paes: a memetic algorithm for multi-objective optimization. In: Proceedings of the 2000 Congress on Evolutionary Computation, vol. 1, pp. 325–332. IEEE (2000)
44. Krishnanand, K., Ghose, D.: Detection of multiple source locations using a glowworm metaphor with applications to collective robotics. In: Proceedings 2005 IEEE Swarm Intelligence Symposium. SIS 2005, pp. 84–91. IEEE (2005)
45. Kumar, S., Chaturvedi, D.: Tuning of particle swarm optimization parameter using fuzzy logic. In: 2011 International Conference on Communication Systems and Network Technologies (CSNT), pp. 174–179. IEEE (2011)

46. Lim, D., Ong, Y.S., Lim, M.H., Jin, Y.: Single/multi-objective inverse robust evolutionary design methodology in the presence of uncertainty, pp. 437–456 (2007)
47. Lim, K.K., Ong, Y.S., Lim, M.H., Chen, X., Agarwal, A.: Hybrid ant colony algorithms for path planning in sparse graphs. Soft Comput. **12**(10), 981–994 (2008)
48. Lim, M., Xu, Y.: Application of hybrid genetic algorithm in supply chain management. Int. J. Comput. Syst. Signals. Special issue on Multi-objective Evolution: Theory and Applications **6**(1) (2005)
49. Lim, M.H., Gustafson, S., Krasnogor, N., Ong, Y.S.: Editorial to the first issue. Memet. Comput. **1**, 1–2 (2009)
50. Loughlin, D.H., Ranjithan, S.R.: The neighborhood constraint method: a genetic algorithm-based multi-objective optimization technique. In: ICGA, pp. 666–673 (1997)
51. McMullen, P.R.: An ant colony optimization approach to addressing a JIT sequencing problem with multiple objectives. Artif. Intell. Eng. **15**(3), 309–317 (2001)
52. Moscato, P., et al.: On evolution, search, optimization, genetic algorithms and martial arts: towards memetic algorithms. Caltech concurrent computation program, C3P Report, vol. 826 (1989)
53. Mostaghim, S., Teich, J.: Strategies for finding good local guides in multi-objective particle swarm optimization (mopso). In: Proceedings of the 2003 IEEE Swarm Intelligence Symposium. SIS 2003, pp. 26–33. IEEE (2003)
54. Mühlenbein, H., Schlierkamp-Voosen, D.: Predictive models for the breeder genetic algorithm I. Continuous parameter optimization. Evol. Comput. **1**(1), 25–49 (1993)
55. Müller, S., Airaghi, S., Marchetto, J., Koumoutsakos, P.: Optimization algorithms based on a model of bacterial chemotaxis. In: Proceedings of 6th International Conference on Simulation of Adaptive Behavior: From Animals to Animats, SAB 2000 Proc. Suppl. Citeseer (2000) Proceedings supplement Citeseer
56. Ong, Y., Keane, A.: A domain knowledge based search advisor for design problem solving environments. Eng. Appl. Artif. Intell. **15**(1), 105–116 (2002)
57. Ong, Y.S., Lim, M.H., Zhu, N., Wong, K.W.: Classification of adaptive memetic algorithms: a comparative study. IEEE Trans. Syst. Man Cybern. Part B (Cybern.) **36**(1), 141–152 (2006)
58. Ong, Y.S., Nair, P.B., Keane, A.J.: Evolutionary optimization of computationally expensive problems via surrogate modeling. AIAA J. **41**(4), 687–696 (2003)
59. Ong, Y.S., Nair, P.B., Lum, K.Y.: Max-min surrogate-assisted evolutionary algorithm for robust design. IEEE Trans. Evol. Comput. **10**(4), 392–404 (2006)
60. Passino, K.M.: Biomimicry of bacterial foraging for distributed optimization and control. IEEE Control Syst. **22**(3), 52–67 (2002)
61. Poloni, C.: Hybrid GA for multi-objective aerodynamic shape optimization. pp. 397–415. Wiley, New York (1995)
62. Price, K.V.: Differential evolution versus the functions of the 2/sup nd/ICEO. In: IEEE International Conference on Evolutionary Computation, pp. 153–157. IEEE (1997)
63. Price, K.V.: An introduction to differential evolution. New ideas in optimization, pp. 79–108 (1999)
64. Price, K.V., Storn, R.M., Lampinen, J.A.: Differential Evolution A Practical Approach to Global Optimization. Springer (2005)
65. Qin, A.K., Huang, V.L., Suganthan, P.N.: Differential evolution algorithm with strategy adaptation for global numerical optimization. IEEE Trans. Evol. Comput. **13**(2), 398–417 (2009)
66. Qiu, C., Wang, C., Zuo, X.: A novel multi-objective particle swarm optimization with k-means based global best selection strategy. Int. J. Comput. Intell. Syst. **6**(5), 822–835 (2013)
67. Dawkins, R.: The Selfish Gene. Oxford University Press (1976)
68. Robič, T., Filipič, B.: Differential evolution for multi-objective optimization. In: Evolutionary Multi-Criterion Optimization, pp. 520–533. Springer (2005)
69. Rosenberg, R.S.: Simulation of genetic populations with biochemical properties. Ph.D. thesis, University of Michigan, Ann Arbor (1967)
70. Rudolph, G.: Convergence analysis of canonical genetic algorithms. IEEE Trans. Neural Netw. **5**(1), 96–101 (1994)

71. Sasaki, D., Morikawa, M., Obayashi, S., Nakahashi, K.: Aerodynamic shape optimization of supersonic wings by adaptive range multi-objective genetic algorithms. In: International Conference on Evolutionary Multi-Criterion Optimization, pp. 639–652. Springer (2001)
72. Schaffer, J.D.: Some experiments in machine learning using vector evaluated genetic algorithms. Ph.D. thesis, Vanderbilt University, Nashville, TN (USA) (1984)
73. Shi, Y., Eberhart, R.C.: Fuzzy adaptive particle swarm optimization 1, 101–106 (2001)
74. Srinivas, N., Deb, K.: Muiltiobjective optimization using nondominated sorting in genetic algorithms. Evol. Comput. 2(3), 221–248 (1994)
75. Stender, J.: Parallel Genetic Algorithms: Theory and Applications, vol. 14. IOS press (1993)
76. Storn, R.: Differential Evolution Research—Trends and Open Questions. Springer (2008)
77. Storn, R., Price, K.: Differential evolution-a simple and efficient adaptive scheme for global optimization over continuous spaces. Berkeley Int. Comput. Sci. Inst. 3 (1995)
78. Storn, R., Price, K.: Minimizing the real functions of the ICEC 1996 contest by differential evolution. In: Proceedings of IEEE International Conference on Evolutionary Computation, pp. 842–844. IEEE (1996)
79. Storn, R., Price, K.: Differential evolution—a simple and efficient heuristic for global optimization over continuous spaces. J. Glob. Optim. 11(4), 341–359 (1997)
80. Sutton, A.M., Lunacek, M., Whitley, L.D.: Differential evolution and non-separability: using selective pressure to focus search. In: Proceedings of the 9th Annual Conference on Genetic and Evolutionary Computation, pp. 1428–1435. ACM (2007)
81. Tang, J., Lim, M.H., Ong, Y.S.: Diversity-adaptive parallel memetic algorithm for solving large scale combinatorial optimization problems. Soft Comput. A Fus. Found. Methodol. Appl. 11(9), 873–888 (2007)
82. Van Veldhuizen, D.A., Lamont, G.B.: Multi-objective evolutionary algorithms: analyzing the state-of-the-art. Evol. Comput. 8(2), 125–147 (2000)
83. Voigt, H.M.: Soft Genetic Operators in Evolutionary Algorithms, pp. 123–141 (1995)
84. Wolpert, D.H., Macready, W.G.: No free lunch theorems for optimization. IEEE Trans. Evol. Comput. 1(1), 67–82 (1997)
85. Yang, S., Wang, M., et al.: A quantum particle swarm optimization 1, 320–324 (2004)
86. Yang, X.S.: A new metaheuristic bat-inspired algorithm. In: Nature inspired cooperative strategies for optimization (NICSO 2010), pp. 65–74 (2010)
87. Zhang, Q., Li, H.: Moea/d: a multi-objective evolutionary algorithm based on decomposition. IEEE Trans. Evol. Comput. 11(6), 712–731 (2007)
88. Zhang, W., Jin, Y., Li, X., Zhang, X.: A simple way for parameter selection of standard particle swarm optimization. Artif. Intell. Comput. Intell. 436–443 (2011)
89. Zhang, Y., Balochian, S., Agarwal, P., Bhatnagar, V., Housheya, O.J.: Artificial intelligence and its applications. Math. Probl. Eng. (2014)
90. Zhang, Y., Wang, S., Ji, G.: A comprehensive survey on particle swarm optimization algorithm and its applications. Math. Probl. Eng. (2015)
91. Zhu, Z., Ong, Y.S., Zurada, J.M.: Identification of full and partial class relevant genes. IEEE/ACM Trans. Comput. Biol. Bioinf. 7(2), 263–277 (2010)
92. Zitzler, E., Laumanns, M., Thiele, L.: Spea 2: improving the strength pareto evolutionary algorithm for multi-objective optimization. In: Giannakoglou, K., Tsahalis, D., Périaux, J., Papailiou, K., Fogarty, T. (eds.) Evolutionary Methods for Design Optimization and Control with Applications to Industrial Problems, pp. 95–100. CIMNE, Athens (2001)
93. Zitzler, E., Thiele, L.: Multi-objective evolutionary algorithms: a comparative case study and the strength pareto approach. IEEE Trans. Evol. Comput. 3(4), 257–271 (1999)

Chapter 2
Complex Networks

Many systems could be abstracted and get represented by complex networks. These include both natural systems such as gene regulatory networks and man-made systems such as the Internet, e-mail exchange networks, phone call networks, science collaboration networks, and citation networks. The importance of understanding complex networks and their power in explaining complex systems have attracted significant interest that resulted in a surge in the amount of research in this area over the last decade. This research could be categorized into two groups. One group focuses on understanding properties of actual networks, while the second, closely dependent, group attempts to trace the root-causes of the emergence of these properties.

The remainder of this chapter will offer a gentle introduction to a selected sample of network properties that are essential for the materials to be presented in the rest of this book, some key network generation algorithms, and how evolutionary computation as being described in the previous chapter gets coupled with complex networks to understand evolution when get subjected to interactions constrained with the topological structure of the network.

2.1 Network Properties

2.1.1 Basic Concepts and Notations of Networks

A graph is a collection of vertices (nodes) connected through edges (links). The area of graph theory studies these topological structures on a high level of abstraction that does not take into account the functions these nodes and/or links perform. While the word "network" is sometimes used as a synonym to a "graph", it is more appropriate to see complex networks as a field that studies graph and the functions and roles the

© Springer Nature Switzerland AG 2019
J. Liu et al., *Evolutionary Computation and Complex Networks*,
https://doi.org/10.1007/978-3-319-60000-0_2

nodes and links perform within this graph. In this chapter, we will use the concept
of a network and leave it to the context to differentiate between the graph structure
and the functions of the graph.

A **graph** G is a pair $G = \langle V, E \rangle$, where $V = \{v_1, v_2, \ldots, v_N\}$ is a set of **vertices**
(nodes) and $E \subseteq V \times V = \{(i, j) \mid v_i, v_j \in V \text{ and } i \neq j\}$. The edges E is a set of
edges that connect these vertices together.

A **network** N is a graph over some set V in which either the nodes or the edges
have attributes associated with them. The **adjacency matrix** A_{ij} of a graph G is a
matrix in which the rows and columns correspond to the vertices and the nonzero
entries correspond to edges in G (Fig. 2.1). For a directed graph, A_{ij} is defined by

$$a_{ij} = \begin{cases} 1 & g_{ij} \in G \\ 0 & otherwise \end{cases} \tag{2.1}$$

Each matrix M is associated with a directed graph G_M, defined by

$$G_M = \{(i, j) \mid m_{ij} \neq 0\} \tag{2.2}$$

The number of edges connected to a node in an undirected graph is called the
degree of that node. Given the directed nature of a directed graph, this metric is
split into the **in-degree**, representing nodes pointing to the node of interest, and the
out-degree, representing nodes that with arcs emitting from the node of interest. For
N nodes, if $a_{i,j}$ is the link between v_i and v_j, $i, j = 1, 2, \ldots, N$; that is, $a_{i,j} = 1$
when edge (i, j) exists, and $a_{i,j} = 0$ otherwise. The **degree** k_i, of vertex v_i is:

$$k_i = \sum_{j=1}^{N} a_{i,j} \tag{2.3}$$

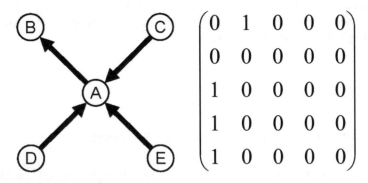

Fig. 2.1 Representations of a simple network. LEFT: A network shown as a diagram. A–E are the
vertices (nodes). The in-degree of node A is three, and its out-degree is 1. RIGHT: the adjacency
matrix of the same network. The rows and columns denote the nodes A–E in order. The nonzero
entries denote a (directed) edge from the node denoted by the row to node denoted the column

All nodes within one hop from v_i are called the **nearest neighbors** of v_i. For a node v in an undirected graph $G = \langle V, E \rangle$, the neighborhood, N_v, of v is defined as $N_v = \{w \mid w \in V \text{ and } (w, v) \in E\}$. Given symmetry, the total number of edges M in the network can be calculated as

$$M = \frac{1}{2} \sum_{i=1}^{N} k_i \tag{2.4}$$

A path of length $(n + 1)$ from v_i to v_j, is an ordered n-tuple of nodes $(v_i, w_1, \ldots w_n, v_j)$, where $(w_k, w_k + 1) \in E \; \forall k$.

The length of the shortest path (distance) between v_i and v_j will be denoted by $d_{i,j}$, $d_{i,j} = \infty$ denotes the absence of a path between the two nodes. A graph is **connected** when there is a path between each pair of nodes in the graph; otherwise, it is called disconnected. The characteristic (path length) of a graph G, also known as the **diameter** $D(G)$ is the maximal distance separating a pair of nodes, where

$$D(G) = max\{d(v_i, v_j) \mid v_i, v_j \in V\} \tag{2.5}$$

For a disconnected graph, the diameter is theoretically defined to be infinite. However, it could practically be defined as the diameter of the giant (largest connected) component in the graph.

2.1.2 Network Topology

The variety of networks in phenomena in nature are found to map to a few forms of network topologies. Some of these are shown in Fig. 2.2. Some important topologies are described below.

- *Cycles.*

  When the start node of a path is identical to the end node, the subgraph is said to form a cycle. Formally, a graph $< G, E >$ is a cycle iff

$$E = \{\langle g_i, g_j \rangle \mid i < |g|\} \cup \{\langle g_{|g|}, g_i \rangle\} \tag{2.6}$$

Cycles are important in networks because they are necessary conditions for positive and negative feedback loops. These loops could either amplify flow, as in the case of positive feedback loops, or stabilize flow as in the case of negative ones. Networks could maintain many of these cycles as subnetworks, generating complex rich dynamics.

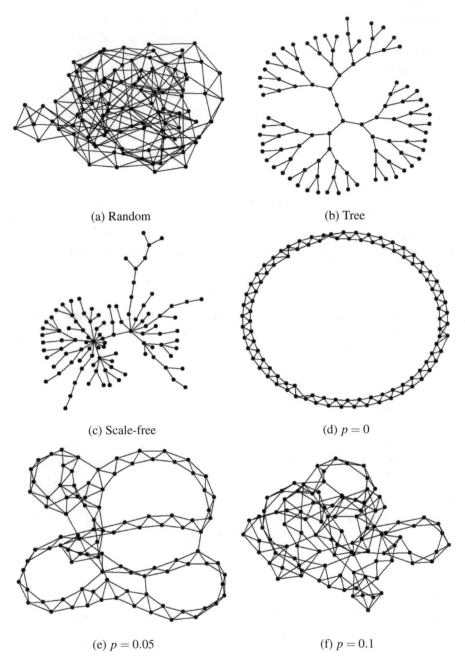

(a) Random   (b) Tree

(c) Scale-free   (d) $p = 0$

(e) $p = 0.05$ (f) $p = 0.1$

Fig. 2.2 Some common network topologies. The bottom line shows small worlds with different probabilities (p) of edges that make long-range connections

- *Trees.*
  The absence of any cycle in a connected undirected graph generate a form of a graph known as trees. Similar to an organization chart, trees are generally important for hierarchical organization of structures and modularization, both assist in managing complexity. However, these forms of structures are fragile with a single breakdown of a link and the graph becomes disconnected. Equally, the addition of any extra nonredundant link would create cycles in an undirected graph.
- *Regular networks.*
  Regular networks usually form when an external process exerts a force on the structure to generate repeated patterns. In human-made networks, these forces could be the designer of a city road system; where a good example could be seen in the rectangular design of Adelaide city.
- *Small worlds.*
  Small-world networks have gained fame from their resemblance to human social networks; the famous six-degree of separation concept, whereby any two persons in the world are only six hops away from each other. These networks are characterized by small geodesic distances and high clustering coefficient.
- *Scale-free networks.*
  When the nodal degrees in a network decays from very high to low in an exponential manner similar to a power law distribution, the network is called scale-free. More specifically, given a node of degree r, and two constant scaling factors a and k, the probability of r is:

$$p(r) = ar^{-k} \tag{2.7}$$

These networks grow according to the famous say the rich gets richer, where new nodes prefer existing nodes with high degrees. A good example of these networks is the WorldWide Web. The complexity of these networks could be measured in multiple folds, but a common metric is *edge density*, which sits on the premise that networks with higher edge density allow for more flow and richer dynamics to occur.

Given G, an undirected simple graph G with N nodes, the maximum possible number of distinct edges is $1/2N(N-1)$. Edge density $E(G)$ is the proportion of the true number of edges in the network, E, and that maximum; that is,

$$E(G) = \frac{2E}{N(N-1)} \tag{2.8}$$

2.1.3 Modularity

The division of a system into smaller self-contained components is known as encapsulation. This phenomenon enables natural and artificial systems to manage complexity. Encapsulation generates modular structure that allows the system to grow and complexity to increase without increasing the complexity of the process underneath "the system's" growth itself.

Software engineering is a good example, with innovations such as component programming or the classic object-oriented languages. A large piece of code could be structured into smaller reusable pieces of code. Size and functionality of the code increases, while there exists underneath these massive pieces of software a smaller finite building blocks that get used over and over when and where needed.

There is an ample of examples on the use of modularity in nature, from the growth of trees to the growth of an organism through modular genome structures [9]. In computational models of evolution, modularity is seen to promote evolvability [7, 19, 20].

Graph theory and networks see modularity in terms of structure. A set of nodes with more connectedness among themselves than others emerge as modules as shown in Fig. 2.3. Modules get normally repeated in a structure but equally could stand on their own holding important roles within the overall structure without reoccurrence.

In a network with N nodes, given a neighboring pair of nodes i and j, if C_{ij} is the number of connections they share and A_i and A_j represent the number of other neighbors of i and j, respectively, a local measure for the relative richness of internal connections for their joint neighborhood could be defined as:

$$m_{ij} = C_{ij}/(A_i + A_j + C_{ij})$$

Averaging m_{ij} over all pairs of nodes i and j provides the global measure $0 \geq M \leq 1$:

$$M = \frac{1}{N(N-1)} \sum_{\substack{i=1}}^{N} \sum_{\substack{j=1 \\ j \neq i}}^{N} m_{ij} \tag{2.9}$$

The above measure for modularity [17] relies on the ratio of internal connections to external ones. While it can be calculated independently of prior knowledge on

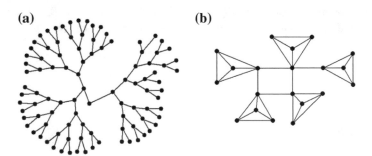

(a)   **(b)**

Fig. 2.3 Two examples of simple modular networks. **a** tree graph (removing any edge breaks off an entire branch of connected nodes) and **b** network consisting of identical modules (each module is a fully connected network of four nodes with only a single edge connecting them to other modules)

which subgraph is a module, it is highly correlated with the edge density metric; thus, it does not distinguish the neighborhood of a node that falls within the module and the one that falls outside the module.

An alternative measure in the case when modules are known relies on using our prior knowledge of the modules to partition the graph. We first need to define e_{ij} to be the expected number of edges between two nodes i and j, where k_i and k_j are the degrees of the two nodes, respectively. In a network with N nodes and M edges, e_{ij} is calculated as follows:

$$e_{ij} = \frac{k_i k_j}{2M} \tag{2.10}$$

Given $C = c_i$, $i = 1 \ldots p$ defined as a partition over all nodes in the network, and A_{ij} as the network's adjacency matrix, the modularity measure $-1 >= Q \leq 1$ is defined as:

$$Q = \frac{1}{2M} \sum_{i,j,i \neq j}^{p} [A_{ij} - e_{ij}] \tag{2.11}$$

2.1.4 Power Law Degree Distribution

When grouping the nodes in a graph according to their degree, calculating the frequency of nodes for each unique degree in the graph, dividing each frequency by the total number of nodes in the graph, and plotting that, the resultant figure represents the probability density function for the degree distribution of that graph. Corresponding to each degree on the x-axis, the corresponding y-axis value represents the probability that a node chosen at random from the network will have this degree. We denote this distribution as $P(k)$, where k is a degree.

In a random graph, it is expected that $P(k)$ is a Poisson distribution given the unbiased attachment strategy associated with the addition of a node to the graph and the bias generated from sequencing the attachment of nodes. In this case, $P(k)$ will asymptotically approach the average degree, $\langle k \rangle$, of the network, and the distribution will have a peak at this value.

Preferential attachment was introduced as an alternate strategy that biases the probability that a newcomer will attach itself to an existing node in a graph to the degree of that node. The basic idea is the richer gets richer and hubs are better magnets than nonhub nodes.

This attachment strategy causes the degree distribution to deviate from Poisson into a power law. This form of a network is known as "scale-free network" and is attributed to Barabasi [1, 2].

The degree distribution of a power law network would have a long tail as the network size increases. Given γ as the power law exponent, the distribution is defined as:

$$P(k) \propto k^{-\gamma} \tag{2.12}$$

2.1.5 Clustering Coefficient

Social networks tend to have a significant amount of transitivity; that is, the friend of my friend is also my friend phenomenon. These triangles form cliques of well-connected groups. Wasserman [5, 21] described this as the "fraction of transitive triples", while Watts [22] formally developed it into what is known today as the "clustering coefficient".

For a given vertex v_i with k_i degree, if the immediate neighbors of v_i form a clique, we expect this clique to have $\frac{k_i(k_i-1)}{2}$ edges. The clustering coefficient, C_i, of node v_i is calculated using E_i, the actual number of edges that exist in the clique as follows:

$$C_i = \frac{2E_i}{k_i(k_i-1)} \qquad (2.13)$$

For the overall graph, the clustering coefficient, C, is the average of the clustering coefficients of all nodes. Equivalently, we can take the ratio between the number of triangles, or trios of complete subgraphs, and the number of connected triples, where a trio is connected but not necessarily forming a complete graph, the metric is calculated as [14]

$$C = \frac{3 \times \text{Number of triangles}}{\text{Number of connected triples}} \qquad (2.14)$$

A random graph exhibits a smaller clustering coefficient of $C = \langle k \rangle / N$, while real-world social networks would normally have a significantly larger cluster coefficient.

2.1.6 Small-World Networks

Social networks have been characterized with the "six degrees of separation" phenomenon [18], which was the basis for the small-world phenomenon. The basic idea is that when selecting any two nodes at random from a large network, we can always find a path between these two nodes with a maximum of six other nodes.

While initially thought as a social phenomenon, it then appeared to be a characteristic for many complex networks that warranted a measure to estimate the degree of which a network follows a small-world phenomenon [8].

Given C and C_{rand} to be the clustering coefficients of the network and a random graph respectively, and L and L_{rand} to be the characteristic path lengths of the network and a random graph, respectively, where

$$L = \frac{1}{N}\sum_{i=1}^{N} L_i = \frac{1}{N}\sum_{i=1}^{N} \frac{\sum_{j=1,j\neq i}^{N} d_{ij}}{N-1} \qquad (2.15)$$

the small-world phenomenon could be measured using S, typically has a value much larger than 1, as the trade-off between high local clustering and short path length:

$$S = \frac{C/C_{\text{rand}}}{L/L_{\text{rand}}} \tag{2.16}$$

2.1.7 Assortative Mixing

"Assortative mixing" represents the degree similarity of adjacent nodes in a graph. If it is more common that high-degree nodes are connected to similar high-degree nodes and low-degree nodes are connected to similar low-degree nodes than it is that high-degree nodes are connected to low-degree nodes, the network will tend to have higher assortative mixing measure [12]. For a node v_i, the metric requires the calculation of the average degree of the neighbors of v_i as follows [15]:

$$k_{nn,i} = \frac{\sum_{j=1}^{N} a_{ij} k_j}{k_i} \tag{2.17}$$

For an undirected graph, the Pearson correlation coefficient between the degrees of adjacent nodes approximate assortative as follows:

$$r = \frac{M^{-1} \sum_{i=2}^{N} \sum_{j=1}^{i} a_{ij} k_i k_j - [M^{-1} \sum_{i=2}^{N} \sum_{j=1}^{i} \frac{1}{2} a_{ij} (k_i + k_j)]^2}{M^{-1} \sum_{i=2}^{N} \sum_{j=1}^{i} \frac{1}{2} a_{ij} (k_i^2 + k_j^2) - [M^{-1} \sum_{i=2}^{N} \sum_{j=1}^{i} \frac{1}{2} a_{ij} (k_i + k_j)]^2} \tag{2.18}$$

Due to this measure's reliance on the correlation coefficient, it is always between -1 and 1; ranging from highly negatively correlated (high degree nodes connect to low-degree nodes) to highly positively correlated, where high and low-degree nodes connect to high and low-degree nodes, respectively.

2.1.8 Community Structure

A community is defined by the density of connections within a subgraph relative to the density of connections of the subgraph to the rest of the network. These forms of dense clusters naturally occur in social networks, biological self-regulatory networks, and computer networks.

The detection of community structures is a very active area of research. Newman et al. [13] contributed a measure for modularity: Q. Given a network with m disjoint communities, e_{ii}, the fraction of edges that fall within community i, and a_i

is the fraction of all ends of edges that are attached to vertices in community i, the modularity measure Q is calculated by

$$q = \sum_{i=1}^{m}(e_{ii} - a_i^2) \qquad (2.19)$$

Typical practical values for Q would be between 0.3 and 0.7 indicating strong community structures present in a network. The higher the value of Q, the less likely that the links among the nodes of the subgraph forming the community exist because of a random chance.

Given two nodes v_i and v_j belonging to communities c_i and c_j, respectively, the δ-function $\delta(c_i, c_j)$ needs to be set to 1 when the two communities are identical and 0, otherwise, a second measure of modularity that was proposed by Clauset et al. [4] as follows:

$$q = \frac{1}{2M} \sum_{i=1}^{n} \sum_{j=1}^{n} [a_{ij} - \frac{k_i k_j}{2M}]\delta(c_i, c_j) \qquad (2.20)$$

2.2   Network Generation Methods

Studying real-world networks is limited with the finite sources of data. Taking the WorldWide Web (WWW) as an example, there is only one WWW network. We could possibly sample subnetworks out of it, but still the problem remains the same; that is, the data is limiting to study a wide variety of questions in a scientific manner that requires testing on tens, hundreds, or thousands of networks.

Researchers resorted to developing network generation algorithms. These algorithms take as inputs the characteristics of the network of interest and output randomly generated networks with these characteristics. The complexity of these algorithms range from simple ones, where a single characteristic is used as in the example of generating networks with a specific degree distribution, to very complex ones requiring advanced optimization algorithms to find the closest network match that trade-off many characteristics.

In this section, we will review some of the basic network generation models. They mostly fall in the first category above of simple network generation algorithms. Their simplicity comes with the advantage that they are mostly very efficient and generate their target networks using a sequential method that does not require heavy optimization steps.

2.2.1 Erdös-Rényi Model

Erdös-Rényi [5] produced one of the earliest algorithms for generating random graphs. In their first attempt to generate a random graph with M edges selected from all possible $N(N-1)/2$ edges of a network with N nodes, they proposed an iterative network generation algorithm that works as follows. Starting with N nodes, an edge is created between a random pair of nodes that gets selected without replacement using a uniform distribution. The process continues until the target number of edges M is reached.

With no restriction on M, there is no guarantee that the resultant network is connected. However, this problem could be easily fixed. Given that a network with N nodes require a minimum of $N-1$ edges to be connected, after selecting the first edge to add to the network, the first $N-2$ randomly selected edges after that could be selected with a condition that one of the nodes in the selected edge already exists in the set of nodes of existing edges, while the second node is new. This will ensure a connected graph.

Moreover, the unbiased selection of edges entails that the method can only generate a limited form of networks. Moreover, the method is not guided with any social or biological process for network generation. As such, it would generate theoretical networks that do not resemble any real-world characteristics.

Nevertheless, the above method has attracted significant amount of attention and has been analyzed mathematically extensively.

2.2.2 Small-World Network Generation Model

The six-degrees of freedom discussed above and their associated small-world networks required a network generation algorithm that can generate networks with small diameter and high clustering. The Watts–Strogatz model [22] was the first to be able to generate this type of networks. The advantage of their model is that it only has a single parameter that allows the generated network to sit anywhere between an ordered finite-dimensional lattice and a random graph.

The algorithm starts by ordering the N nodes into a ring lattice structure. Each of the N nodes gets connected to its first K neighbors by balancing half of the connections on each side. To ensure that the network is sparse, the following condition needs to be enforced: $N \gg K \gg ln(N) \gg 1$.

Once the ring lattice structure is formed, the algorithm iterates by randomly rewriting each edge in the lattice with the control parameter $1 \leq p \leq 0$, which represents the probability of rewiring to take place. The rewiring can only be done if it does not lead to a self-connection or duplicate edges. The rewiring algorithm generates $pNK/2$ long bridges between different neighborhoods, thus reducing the diameter of the network. It is important to notice that at the same time it reduces the diameter, it could also reduce the clustering coefficient; hence, it is important to choose

p carefully to balance the reduction needed in the diameter without sacrificing the clustering coefficient needed in the generated graph.

The Watts-Strogatz model is rooted in social systems. However, the degree distribution of the resultant networks resembles more a random graph than a scale-free one.

2.2.3   Scale-Free Network Generation Model

Barabási-Albert [1] introduced the first algorithm to generate networks that follow the power law degree distribution. The phase algorithm follows a growth phase while enforcing a preferential attachment strategy. The growth phase ensures that the generated network is connected. A small number of nodes ($m_0 \ll N$) are first selected. At each growth step, a new node with ($m \leq m_0$) edges connects to existing m different nodes. The attachment strategy biases these connections to existing nodes with higher degrees. This is achieved probabilistically using the probability ($\Pi(k_i)$) that an existing node will attract the newly added node to connect to it. If an existing node v_i has a degree k_i, the probability of this node to be selected for connection to the newly added node is:

$$\Pi(k_i) = \frac{k_i}{\sum_{j=1}^{N} k_j} \tag{2.21}$$

As the network grows, highly connected nodes get exponentially higher chances to connect to newly added nodes. This biased attachment method generates long power law tails as N increases. Other variations of Barabási's preferential attachment were introduced in the literature including the "copying model" [10],"winner does not take all model" [16], "forest fire model" [11], and "random surfer model" [3].

2.2.4   Community Network Generation Model

This class of network generation algorithms focuses on generating networks with community structures in addition to the previous classic network properties such as power law distribution.

The first algorithm by González et al. [6] uses a set of moving particles, called mobile agents, to represent nodal movements in space. Each time a particle establishes contact with another through collisions, a connection between the two nodes is created.

The algorithm starts with N moving particles occupying a circular space with radius r and moving with the same velocity, v_0 but with randomly initialized directions. The particles move in a square with width L. Periodic boundary conditions are used, where when particles reach one side of the environment, they appear from the opposite side. The size of the environment needs to be much larger than the number of particles to ensure low density (ρ) such that $\rho \equiv N/L^2$.

When particles collide and a link is created between their two perspective nodes, the particles change directions at random creating continuous flips between drifting and diffusing behaviors. Their velocities are updated

The location $\mathbf{x}_i$ of agent i is updated at each time step Δt using the following equation

$$\mathbf{x}_i(t+1) = \mathbf{x}_i(t)\mathbf{v}_i(t)\Delta t \qquad (2.22)$$

Following collisions, for each agent i, the velocity modulus are updated in proportion to their degree k_i as follows:

$$|\mathbf{v}_i(t)| = v_0 + \bar{v}k_i(t) \qquad (2.23)$$

where $\bar{v}$ is a constant with a unit velocity.

The age of agent i, A_i, is initialized at random from the interval $[0, T_l]$, then gets updated as:

$$A_i(t+1) = A_i(t) + \Delta t \qquad (2.24)$$

When the maximum bound of the interval is reached; that is, $A_i = T_l$, the corresponding agent i exits the system. Each time an agent exits the system, all their properties and links get deleted. A new agent is created and takes the position of the old one.

Networks created using this procedure can resemble real-world networks in their degree distribution, clustering coefficient, shortest path length, and community structures.

References

1. Barabási, A.L., Albert, R.: Emergence of scaling in random networks (1999)
2. Barabási, A.L., Albert, R., Jeong, H.: Mean-field theory for scale-free random networks (1999)
3. Blum, A., Chan, T.H., Rwebangira, M.R.: A random-surfer web-graph model. In: 2006 Proceedings of the Third Workshop on Analytic Algorithmics and Combinatorics (ANALCO), pp. 238–246. SIAM (2006)
4. Clauset, A., Newman, M.E., Moore, C.: Finding community structure in very large networks. Phys. Rev. E **70**(6) (2004)
5. Erdős, P., Rényi, A.: On the evolution of random graphs. Magyar Tud. Akad. Mat. Kutató Int. Közl **5**, 17–61 (1960)
6. González, M.C., Lind, P.G., Herrmann, H.J.: System of mobile agents to model social networks. Phys. Rev. Lett. **96**(8) (2006)
7. Hansen, T.F.: Is modularity necessary for evolvability? Remarks on the relationship between pleiotropy and evolvability. Biosystems **69**(2), 83–94 (2003)
8. Humphries, M.D., Gurney, K.: Network small-world-ness: a quantitative method for determining canonical network equivalence. PloS one **3**(4) (2008). (DOI e0002051)

9. Jacob, F., Monod, J.: Genetic regulatory mechanisms in the synthesis of proteins. J. Mol. Biol. **3**(3), 318–356 (1961)
10. Kumar, R., Raghavan, P., Rajagopalan, S., Sivakumar, D., Tomkins, A., Upfal, E.: Stochastic models for the web graph. In: Proceedings of 41st Annual Symposium on Foundations of Computer Science, 2000, pp. 57–65. IEEE (2000)
11. Leskovec, J., Kleinberg, J., Faloutsos, C.: Graphs over time: densification laws, shrinking diameters and possible explanations. In: Proceedings of the Eleventh ACM SIGKDD International Conference on Knowledge Discovery in Data Mining, pp. 177–187. ACM (2005)
12. Newman, M.E.: Mixing patterns in networks. Phys. Rev. E **67**(2) (2003)
13. Newman, M.E., Girvan, M.: Finding and evaluating community structure in networks. Phys. Rev. E **69**(2) (2004)
14. Newman, M.E., Strogatz, S.H., Watts, D.J.: Random graphs with arbitrary degree distributions and their applications. Phys. Rev. E **64**(2) (2001)
15. Pastor-Satorras, R., Vázquez, A., Vespignani, A.: Dynamical and correlation properties of the internet. Phys. Rev. Lett. **87**(25) (2001). (DOI 258701)
16. Pennock, D.M., Flake, G.W., Lawrence, S., Glover, E.J., Giles, C.L.: Winners don't take all: characterizing the competition for links on the web. Proc. Natl. Acad. Sci. **99**(8), 5207–5211 (2002)
17. Pimm, S.L.: Food Webs, pp. 1–11 (1982)
18. Travers, J., Milgram, S.: The small world problem. Phychol. Today **1**(1), 61–67 (1967)
19. Wagner, A.: Does evolutionary plasticity evolve? Evolution **50**(3), 1008–1023 (1996)
20. Wagner, G.P., Altenberg, L.: Perspective: complex adaptations and the evolution of evolvability. Evolution **50**(3), 967–976 (1996)
21. Wasserman, S., Faust, K.: Social Network Analysis: Methods and Applications, vol. 8 (1994)
22. Watts, D.J., Strogatz, S.H.: Collective dynamics of 'small-world' networks. Nature **393**(6684), 440 (1998)

Part II
Complex Networks for Evolutionary Algorithms

Chapter 3
Problem Difficulty Analysis Based on Complex Networks

EAs have recently gained increasing interest because they are suitable for solving complex and ill-defined problems. EAs have been successfully applied to the fields of numerical optimization, constraint satisfaction problems, data mining, neural networks, and many other engineering problems. Intrinsically, EAs are still randomized search heuristics, that is, a kind of black-box algorithms. Therefore, one of the major challenges in the field of EAs is to characterize which kinds of problems are easy for a given algorithm to solve and which are not [14], namely, problem difficulty prediction. As a result, many researchers have sought to analyze and predict the behavior of EAs in different domains. The major tool used in problem difficulty prediction is fitness landscapes. Since fitness landscapes are obviously related to networks or graphs, some analysis about problem difficulty from the viewpoint of networks have been conducted.

3.1 Fitness Landscapes

The concept of a fitness landscape was introduced in theoretical genetics [29] as a way to visualize evolutionary dynamics and has been proven to be very powerful in evolutionary theory. Fitness landscapes were connected to EAs via a neighborhood structure based on operators used in EAs, which highlights the association between search spaces and fitness spaces. Here, we consider fitness landscapes only for combinatorial optimization problems. Formally, a fitness landscape $\mathbf{L}$ is defined by a tuple of three components:

$$\mathbf{L} = (\mathbf{S}, \mathbf{f}, \mathbf{N}) \tag{3.1}$$

where $\mathbf{S}$ is the set of candidate solutions in search spaces; $\mathbf{f} : \mathbf{S} \to \mathbf{R}$ is a fitness function, which assigns a numeric value to each candidate solution; and $\mathbf{N}$ is a neighborhood structure defined over $\mathbf{S}$ as follows:

$$\forall s \in \mathbf{S}, \mathbf{N}(s) = \{y \in \mathbf{S} \mid P(y \in \mathbf{operator}(s)) > 0\} \tag{3.2}$$

© Springer Nature Switzerland AG 2019
J. Liu et al., *Evolutionary Computation and Complex Networks*,
https://doi.org/10.1007/978-3-319-60000-0_3

where $P(e)$ denotes the probability of the event e, and **operator**(s) denotes the set of candidate solutions that can be obtained by performing **operator** on s. Clearly, with the above neighborhood structure, the neighbors of a candidate solution s are a set of candidate solutions that can be transformed from s by performing **operator** on it. Since EAs are actually determined by various operators, the above neighborhood structure defined on operators reflects the features of different EAs, and correspondingly, these features can be reflected on the resulted fitness landscapes.

3.2 Network-Based Problem Difficulty Analysis

With the property of neighborhood structure in mind, each fitness landscape actually forms a network. In such a network, each node corresponds to a point in the search space, and each edge connects one point in the search space to one of its neighbors. The fitness values can be viewed as the weight of each node. Therefore, different problems correspond to different networks, and when EAs solve problems, they are actually navigating different networks. From this viewpoint, problem difficulty can be predicted by analyzing the features of corresponding fitness landscape networks. Although it is well-known that fitness landscapes are related to networks or graphs, very few predictive measures have been proposed based on the features of the networks or graphs. However, some analysis about problem difficulty from the viewpoint of networks have been conducted.

Based on the observation that fitness landscapes are actually networks, Ochoa et al. [17, 23] proposed a network characterization of combinatorial fitness landscapes by adapting the notion of inherent networks proposed for energy surfaces [8]. They introduced a network-based model that abstracts many details of the underlying landscape and compresses the landscape information into a weighted, oriented graph which they call the local optima network (LON). The vertices of this graph are the local optima of the given fitness landscape, while the arcs are transition probabilities between local optima basins. Then, they attempted to study fitness landscapes and problem difficulty using network analysis techniques. They used the well-known family of NK (where N stands for the chromosome length and K for the number of gene epistatic interactions within the chromosome) landscapes as an example, and exhaustively extracted LONs on NK landscape instances. This work is the first attempt at using network analysis techniques in connection with the study of fitness landscapes and problem difficulty.

Following this direction, a series of work about LONs have been conducted. In [25], Verel et al. extended the above LON formalism to neutral fitness landscapes, which are common in difficult combinatorial search spaces. Their study was based on two neutral variants of the well-known NK family of landscapes. By using these two NK variants, probabilistic (NK_p), and quantified NK (NK_q), in which the amount of neutrality can be tuned by a parameter, they showed that their new definitions of the optima networks and the associated basins are consistent with the previous definitions for the nonneutral case. Moreover, their empirical study and statistical analysis show that the features of neutral landscapes interpolate smoothly between landscapes with maximum neutrality and nonneutral ones. They found some unknown structural differences between the two studied families of neutral landscapes. But overall, the network features studied confirmed that neutrality, in landscapes with percolating neutral networks, may enhance heuristic search.

In [18], Ochoa et al. extended LONs to incorporate a first-improvement (greedy-ascent) hill-climbing algorithm, instead of a best-improvement (steepest-ascent) one, for the definition and extraction of the basins of attraction of the landscape optima. A statistical analysis comparing best- and first-improvement network models for a set of NK landscapes was presented and discussed. The results suggested structural differences between the two models with respect to both the network connectivity, and the nature of the basins of attraction. The impact of these differences in the behavior of search heuristics based on first- and best-improvement local search was discussed.

LONs have also been used to analyze the landscapes of some classic combinatorial optimization problems. In [6], Daolio et al. used LONs to conduct a thorough analysis of two types of instances of the quadratic assignment problem (QAP). The LONs extracted from the so- called uniform and real-like QAP instances showed features clearly distinguishing these two types of instances. Apart from a clear confirmation that the search difficulty increases with the problem dimension, the analysis provided new confirming evidence explaining why the real-like instances are easier to solve exactly using heuristic search, while the uniform instances are easier to solve approximately.

In [12], Iclanzan et al. pointed out that methods so far used an exhaustive search for extracting local optima networks, limiting their applicability to small problem instances only. To increase scalability, they proposed a new data-driven methodology that approximates the LON from actual runs of search methods. The method enabled the extraction and study of LON corresponding to the various types of instances from the QAP Library, whose search spaces are characterized in terms of local minima connectivity. Their analysis provides a novel view of the unified test bed of QAP combinatorial landscapes used in the literature, revealing qualitative inherent properties that can be used to classify instances and estimate search difficulty.

In [7], Daolio et al. extracted and analyzed LONs for the permutation flow-shop problem. Two widely used move operators for permutation representations, namely swap and insertion, were incorporated into the network landscape model. The performance of a heuristic search algorithm on this problem was also analyzed. In particular, they studied the correlation between LON features and the performance of an iterated local search heuristic. Their analysis revealed that network features can explain and predict problem difficulty. The evidence confirmed the superiority of the insertion operator for this problem.

The community is a well-known property of complex networks and has attracted lots of attentions in the field of complex networks [9]. Recently, the community has also been introduced into the study of LONs. In [5], for two different classes of instances of QAPs, Daolio et al. first produced the LONs, and then applied the approach to detect communities in the LONs. They provided evidence indicating that the two problem instance classes give rise to very different configuration spaces. For the so-called real-like class, the networks possess a clear modular structure, while the optima networks belonging to the class of random uniform instances are less well partitionable into clusters.

In [11], Herrmann et al. conducted an analysis of LONs extracted from fitness landscapes of the Kauffman NK model under iterated local search. Applying the Markov cluster algorithm for community detection to the LONs, they found that the landscapes consist of multiple clusters. This result complemented recent findings in the literature that landscapes can be often decomposed into multiple funnels, which increases their difficulty for iterated local search. Their results suggested that the number of clusters as well as the size of the cluster in which the global optimum is located is correlated to the search difficulty of landscapes, and clusters found by community detection in LONs offer a new way to characterize the multi-funnel structure of fitness landscapes.

3.3   Local Optima Networks of Resource-Constrained Project Scheduling Problems

The resource-constrained project scheduling problem (RCPSP) is a branch of scheduling problems, and the aim is to find a schedule with minimum makespan for each activity in a project subject to precedence and resource constraints. The applications of this problem can be found in many areas, such as construction engineering, software development [20]. So far, many algorithms have been proposed to solve RCPSPs. In [19], Patterson et al. proposed an algorithm guided by the so-called precedence tree, which is a category of branch-and-bound methods. Since the RCPSP is NP-hard, exact methods are not feasible to solve the problem with large-scale efficiency. Many researchers proposed heuristic methods to reduce the computational complexity. In [13], Icmeli et al. developed a tabu search (TS) method with the objective of optimizing the net income of the project. In [10], Hatmann proposed a permutation-based genetic algorithm (GA). In [2], an active list-based simulated annealing (SA) was proposed to solve multi-mode RCPSPs. In recent years, more effective methods have been proposed to solve RCPSPs. In [22], Tian et al. presented an approach with decomposition on time windows for RCPSP. In [1], Anantathanvit et al. proposed radius particle swarm optimization for RCPSPs. In [27], Wang et al. used an organizational evolutionary algorithm to solve multi-mode RCPSPs. In [28], Wang et al. developed an organizational cooperative coevolutionary algorithm for multi-mode RCPSPs. In [30], Yuan et al. proposed a multi-agent genetic algorithm for RCPSPs. In [26], Wang et al. proposed a scale-free-based memetic algorithm to solve RCPSPs.

Although many algorithms have been proposed to solve RCPSPs, little has been done on theoretically analyzing the properties of this problem itself. Czogalla et al. studied the fitness landscapes for RCPSPs in [4]. Cai et al. studied the performance of four representations for RCPSPs using fitness landscape analysis technique [3]. Inspired by the above work on LONs, in this section, we present a study on LONs of the configuration space of RCPSPs. Usually, there are three decoding methods for RCPSPs, namely forward, backward, and hybrid (forward–backward–forward) decodings. Therefore, we first construct LONs of RCPSPs by forward, backward, and hybrid decodings, respectively, and then compare the solution spaces constructed by these three decodings to analyze their performance. In the LONs of RCPSPs, the nodes represent locally optimal configurations and the edges represent the probability of transition between their basins of attractions. Various properties of these LONs for RCPSPs, including community structure, degree distribution, and assortative mixing, are analyzed. The analyses show that LONs for RCPSPs fall into three types, and the hybrid decoding can construct LONs with the best performance.

3.3.1 Resource-Constrained Project Scheduling Problem

The RCPSP deals with scheduling of activities in a project over time and resources. A project can be depicted in a network called activity-on-node (AON) network. Assume that a project consists of an activity set $V = \{0, 1, 2, ..., n, n + 1\}$ where activities 0 and $n + 1$ are dummy beginning and termination activities, respectively. $RR = \{R_1, R_2, .., R_k\}$ is renewable resource set. For activity j ($1 \leq j \leq n$), d_j is the duration and r_{jk} is defined as the kth renewable resource it requires. An instance of RCPSP depicted in an AON network is shown in Fig. 3.1.

From a feasible schedule, we can get a starting time set $S = \{s_0, s_1, ..., s_n, s_{n+1}\}$ and a finishing time set $F = \{f_0, f_1, ..., f_n, f_{n+1}\}$. RCPSPs can be formulated as follows:

$$Minimize(f_n) \tag{3.3}$$

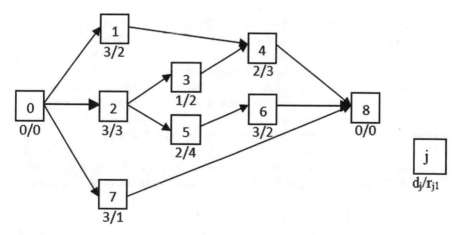

Fig. 3.1 An instance of resource-constrained project scheduling problems

Subject to

$$f_i \leq f_j - d_j \text{ for all } (i, j) \in C \tag{3.4}$$

$$\sum_{i \in S_t} r_{ik} \leq R_k \text{ for } k = 1, 2, ..., |RR| \text{ and } t = 1, 2, ..., f_n \tag{3.5}$$

where C is the precedence constraint set in which (i, j) indicates activity, j is an direct or indirect successor of activity i in the AON network, and S_t is the set of activities that are executed at time t.

Usually, a feasible solution of RCPSPs is a sequence of activities $A = \{a_0, a_1, ..., a_n, a_{n+1}\}$. To transform A into a schedule, three decodings are often used. For an individual A, the forward decoding sets the execution time of a_0 to a_{n+1} as early as possible with precedence and resource constraints satisfied. The backward decoding sets the execution time of a_{n+1} to a_0 as late as possible subject to both constraints. Before describing hybrid decoding, standard representing operation is introduced. Tseng et al. [24] proposed a scheme to transform an individual to its standard representation with the purpose of avoiding the case that two different individuals have the same schedule. In this process, activities with the same starting time are sorted by their index numbers in ascending order, and then hybrid decoding is conducted as follows. First, the forward decoding process is executed. Second, the individual is transformed into its standard representation. Then, the backward decoding is conducted and the corresponding standard representation is obtained. Finally, the individual is decoded by the forward decoding again and a feasible schedule is obtained. For example, for the project given in Fig. 3.1, suppose a feasible individual be $A = (0, 7, 1, 2, 3, 4, 5, 6, 8)$. Then, Fig. 3.2 shows the procedure of hybrid decoding.

3.3.2   Local Optima Networks of RCPSPs

For an instance of RCPSPs with n activities, the fitness landscape [21] contains the following three components: (1) the search space configuration S, which is defined as the permutation of n activities of RCPSPs; (2) the neighborhood structure of activity i labeled as N_i, where $N(i)$ is the set of activities obtained by pairwise exchange operation and where the size of $N(i)$ is $n(n + 1)/2$; (3) the fitness function or objective function of an individual s, which is defined as the makespan of the project for RCPSPs and can be pictured as the height of s in the fitness landscape. In this section, the fitness of an individual is calculated by three different decodings.

Due to precedence constraints, infeasible individuals are abandoned. For the neighborhood structure, an individual generated by pairwise exchange operation is rejected if it breaks precedence constraints. So the actual scale of $N(i)$ is smaller than $n(n - 1)/2$.

Next, the LON is used to capture the fitness landscape of RCPSPs. The nodes are local optima, and edges between them are bidirectional and weighted indicat-

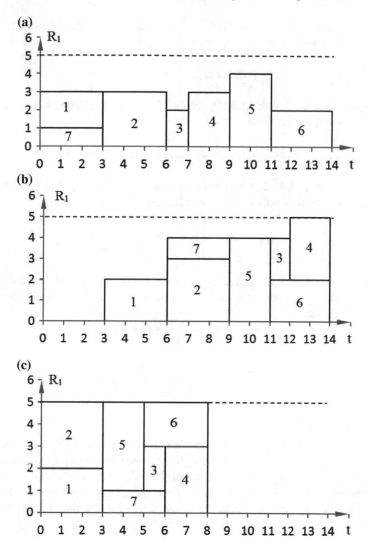

Fig. 3.2 The procedure of hybrid decoding, where **a** forward decoding, **b** backward decoding, and **c** forward decoding

ing the probability of transition. The local optima are obtained by the algorithm called best improvement (BI), which works as follows. Given an individual s and a decoding method, BI searches $N(s)$ and finds the best individual s', then replaces s by s' and repeats the above operation until there is no improvement. For a local optimum s^*, $\forall s \in N(s^*)$, $f(s) > f(s^*)$. We define the basin of attraction to let a local optimum be the representation of a set of solutions.

$$b_{s^*} = \{s \mid BI(s) = s^*\} \tag{3.6}$$

Then, the search space of RCPSPs can be divided into different basins of attractions on different local optima.

$$S = b_1 \cup b_2 \cup \ldots \cup b_n \text{ for } \forall i \neq j, \ b_i \cap b_j = \varnothing \tag{3.7}$$

Finally, the edge between two local optima is considered. The probability of transition between two individuals is calculated as follows:

$$p(s \to s') = \begin{cases} \frac{1}{n(n-1)/2} & s' \in N(s) \\ 0 & \text{otherwise} \end{cases} \tag{3.8}$$

Then, the probability of going from an individual s to the basin of attraction b_i and the probability of going from basin b_i to basin b_j, namely two local optima, are defined as follows:

$$p(s \to b_i) = \sum_{s' \in b_i} p(s \to s') \tag{3.9}$$

$$p(b_i \to b_j) = \frac{1}{b_i} \sum_{s \in b_i} p(s \to b_j) \tag{3.10}$$

where $|b_i|$ is the size of b_i.

Now, the LON of RCPSP is constructed as a graph $G = (S, W)$, where S is the set of local optima and W is the weighting matrix indicating the probability of transition. An element w_{ij} of W is the probability of going from local optimum s_i to local optimum s_j. If $w_{ij}=0$, then the edge between s_i and s_j does not exist. Note that w_{ij} may be different from s_{ij}; hence, the LON of RCPSP is a bidirectional weighted network.

3.3.3 Properties of LONs for RCPSPs

In order to study the properties of LONs for RCPSPs, various problem instances are selected. Considering the largest scale of the search space is equal to $n!$, where n is the number of activities in a project, we choose J10 from the PSPLIB, which has 536 instances and each instance includes 10 activities. Note that J10 is a dataset for a multi-mode resource-constrained project scheduling problem (MRCPSP) in which all the instances in J10 are set to the same mode as the best solution found by now.

3.3.3.1 Topology, Concentration, and Density

In this experiment, three properties of the constructed LONs are analyzed. First, topologies of the three different LONs obtained from the same instance are shown in Fig. 3.3. After analyzing all LONs for the instances in J10, we find that they are connected and the topologies are of the same type. Then, concentration factor c of the LONs is defined as the number of all feasible solutions over the number of local optima, that is, $c = r/r^*$, which is shown in Fig. 3.4. The larger the value of c is, the more concentrated the network is. By decreasing the diversity of solution space, it is easier for an algorithm to find global optima. Finally, the density of LONs is defined as $d=e/E$, where e stands for the number of edges in a LON and E stands for the number of edges in a completely connected network with the same number of nodes.

The decrement of the initial network after BI searches can be evaluated by c. Figure 3.4a shows that there are lots of abnormal values, so we enlarge the normal part and obtain (b). As we can see, the LON constructed by hybrid decoding has the smallest c, followed by forward and finally backward.

Since $c = r/r^*$, a larger value of c means relative less number of local optima, and it is easier to find global optimum in terms of search space. But we realize that the local optima are of certain types, and each type has the same value. The average ratio of global optima in all the nodes is 0.892 for hybrid, 0.827 for forward, and 0.763 for backward decoding. Since the mode is set to the best solution mode, the above ratios are relatively large. From this viewpoint, even LONS constructed by hybrid decoding with the minimum c are still a better solution for the problem.

Then, we calculate the average d of 536 instances, which is 0.009, 0.017, 0.024, for hybrids, forward, and backward, respectively. All LONs are connected and sparse with LONs constructed by hybrid decoding being the sparsest ones.

3.3.3.2 Community Structure

Communities or clusters in networks can be simply defined as being groups of nodes that are strongly connected between them and poorly connected with the rest of the graph. Community detection has been an extraordinary significant aspect in analyzing

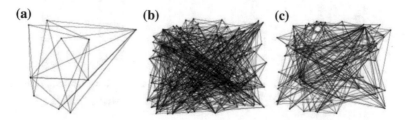

(a)　　　　　　**(b)**　　　　　　**(c)**

Fig. 3.3 Topologies of the three different LONs. **a** LON constructed by the forward decoding, **b** LON constructed by the backward decoding, and **c** LON constructed by the hybrid decoding

properties of networks. Several methods have been proposed to uncover the clusters present in a network. In this experiment, we use the modularity Q to measure the community structure of the constructed LONs. First, in the preprocessing step, we transform the bidirectional weighted graph G into an undirected weighted graph G_u, then we conduct the community detection algorithm proposed by Newman et al. [16]. The results are shown in Fig. 3.5.

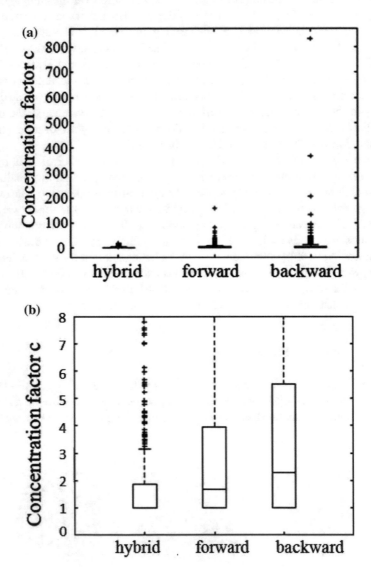

Fig. 3.4 Boxplots of concentration factor c of the three different LONs: **a** original boxplot, and **b** modified boxplot

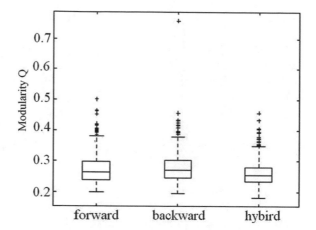

Fig. 3.5 Boxplot of the modularity value Q on the y-axis with respect to the three types of LONs

In general, the higher the value of Q of a partition is, the more significant the community structure is. If $Q > 0.3$ the network can be seen as having the character of community structure. Figure 3.5 shows that community structure does not exist in any of the three types of LONs. This result reveals the diversity of the configuration space for RCPSPs and the difficulty in solving the problem.

3.3.3.3 Degree Distribution

In this experiment, we study the degree distribution of the undirected weighted LONs. Simulated results show that different instances have different degree distributions. Generally, there are four types of degree distribution, as shown in Fig. 3.6, and the degree is close to the median value.

3.3.3.4 Assortative Mixing

In this experiment, we check another important property of complex networks, namely assortative mixing, which indicates tendency for high-degree nodes to associate preferentially with other high-degree nodes. Assortative networks percolate more easily, and they are also more robust to removal of their highest degree nodes [15]. We study the assortative mixing of constructed LONs from the aspect of the average value of nearest neighbors' degree. The relationship between a node's degree and its nearest neighbors' average degree is shown in Fig. 3.7.

The first two figures in Fig. 3.7 show that some instances do not have evident correlation between a node's degree and its nearest neighbors' average degree, while the last two figures indicate that the higher a node's degree is, the higher its nearest neighbors' average degree is. Furthermore, we calculate the ratio of networks for all the 536 instances that have the relationship of assortative mixing. They are 47.95%,

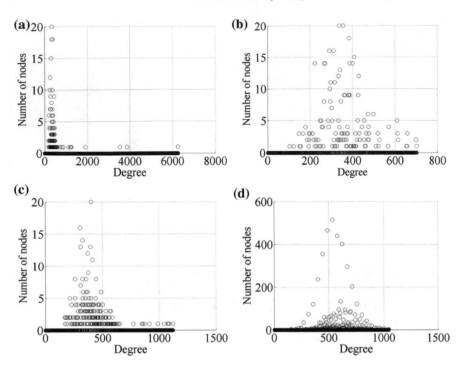

Fig. 3.6 Four types of degree distribution of constructed LONs: **a** degree close to small value and **b**, **c**, **d** degrees close to median value

39.92% and 72.41% for LONs decoded by forward, backward, and hybrid decodings, respectively. The LONs constructed by the hybrid decoding have obviously higher ratio. Hybrid decoding creates LONs robust to intervention and attacks. At this point, they are more similar to social networks.

In summary, the above results indicate that the hybrid decoding can construct LONs with minimum value of c and the largest ratio of global optima. From this point, more local optima can be found using the hybrid decoding method, many of which are global optima. Namely, the hybrid decoding method is more effective in local search. Clear community structure does not exist in any LON discussed above. The plot indicates that the three kinds of LONs are not well separated in terms of Q. As for degree distribution, highly similar features are shown in all of the three types of LONs, with most nodes having a degree of median value. That means most nodes are active during the search. The experimental results on assortative mixing show different features in the three types of LONs. The hybrid decoding method can construct LONs with better performance in terms of assortative mixing.

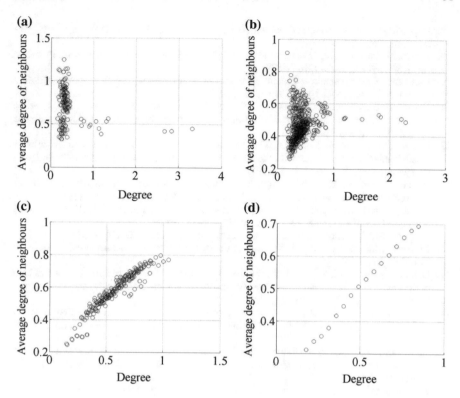

Fig. 3.7 Four types of relationship between a node's degree and its nearest neighbors' average degree appeared in forward, backward, hybrid LONs: **a** nonlinear relationship, **b** nonlinear relationship, **c** linear relationship, and **d** linear relationship

References

1. Anantathanvit, M., Munlin, M.A.: Radius particle swarm optimization for resource constrained project scheduling problem. In: 2013 16th International Conference on Computer and Information Technology (ICCIT), pp. 24–29. IEEE (2014)
2. Boctor, F.F.: A new and efficient heuristic for scheduling projects with resource restrictions and multiple execution modes. Eur. J. Oper. Res. **90**(2), 349–361 (1996)
3. Cai, B., Liu, J.: A study of representations for resource constrained project scheduling problems using fitness distance correlation. In: International Conference on Intelligent Data Engineering and Automated Learning, pp. 218–225. Springer (2013)
4. Czogalla, J., Fink, A.: Fitness landscape analysis for the resource constrained project scheduling problem. LION **3**, 104–118 (2009)
5. Daolio, F., Tomassini, M., Vérel, S., Ochoa, G.: Communities of minima in local optima networks of combinatorial spaces. Phys. A Stat. Mech. Appl. **390**(9), 1684–1694 (2011)
6. Daolio, F., Verel, S., Ochoa, G., Tomassini, M.: Local optima networks of the quadratic assignment problem. In: 2010 IEEE Congress on Evolutionary Computation (CEC), pp. 1–8. IEEE (2010)
7. Daolio, F., Verel, S., Ochoa, G., Tomassini, M.: Local optima networks of the permutation flow-shop problem, pp. 41–52 (2013)

8. Doye, J.P.: Network topology of a potential energy landscape: a static scale-free network. Phys. Rev. Lett. **88**(23), 238701 (2002)
9. Girvan, M., Newman, M.E.: Community structure in social and biological networks. Proc. Nat. Acad. Sci. **99**(12), 7821–7826 (2002)
10. Hartmann, S.: A competitive genetic algorithm for resource-constrained project scheduling. Naval Res. Logist. (NRL) **45**(7), 733–750 (1998)
11. Herrmann, S., Ochoa, G., Rothlauf, F.: Communities of local optima as funnels in fitness landscapes. In: Proceedings of the 2016 on Genetic and Evolutionary Computation Conference, pp. 325–331. ACM (2016)
12. Iclanzan, D., Daolio, F., Tomassini, M.: Data-driven local optima network characterization of qaplib instances. In: Proceedings of the 2014 Annual Conference on Genetic and Evolutionary Computation, pp. 453–460. ACM (2014)
13. Icmeli, O., Erenguc, S.S.: A tabu search procedure for the resource constrained project scheduling problem with discounted cash flows. Comput. Oper. Res. **21**(8), 841–853 (1994)
14. Naudts, B., Kallel, L.: A comparison of predictive measures of problem difficulty in evolutionary algorithms. IEEE Trans. Evol. Comput. **4**(1), 1–15 (2000)
15. Newman, M.E.: Assortative mixing in networks. Phys. Rev. Lett. **89**(20) (2002)
16. Newman, M.E., Girvan, M.: Finding and evaluating community structure in networks. Phys. Rev. E **69**(2) (2004)
17. Ochoa, G., Tomassini, M., Vérel, S., Darabos, C.: A study of nk landscapes' basins and local optima networks. In: Proceedings of the 10th Annual Conference on Genetic and Evolutionary Computation, pp. 555–562. ACM (2008)
18. Ochoa, G., Verel, S., Tomassini, M.: First-improvement versus best-improvement local optima networks of nk landscapes. In: Parallel Problem Solving from Nature, PPSN XI, pp. 104–113 (2010)
19. Patterson, J.H., Sowiski, R., Talbot, F.B., Weglarz, J.: An algorithm for a general class of precedence and resource constrained scheduling problems. In: Sowiski, R., Weglarz, J. (eds.) Advances in Project Scheduling, pp. 3–28. Elsevier, Amsterdam (1989)
20. Rayward-Smith, V.: Project scheduling: recent models, algorithms and applications book review. J. Oper. Res. Soc. **52**(5), 602–603 (2001)
21. Stadler, P.F.: Fitness landscapes. In: Lassig, M., Valleriani, A. (eds.) Biological Evolution and Statistical Physics. Lecture Notes Physics, vol. 585, pp. 187–207. Springer, Heidelberg (2002)
22. Tian, J., Liu, Z., Yu, W.: An approach with decomposition on time windows for resource-constrained project scheduling. In: The 26th Chinese Control and Decision Conference (2014 CCDC), pp. 4897–4903. IEEE (2014)
23. Tomassini, M., Verel, S., Ochoa, G.: Complex-network analysis of combinatorial spaces: the n k landscape case. Phys. Rev. E **78**(6), 066114 (2008)
24. Tseng, L.Y., Chen, S.C.: A hybrid metaheuristic for the resource-constrained project scheduling problem. Eur. J. Oper. Res. **175**(2), 707–721 (2006)
25. Verel, S., Ochoa, G., Tomassini, M.: Local optima networks of nk landscapes with neutrality. IEEE Trans. Evol. Comput. **15**(6), 783–797 (2011)
26. Wang, L., Liu, J.: A scale-free based memetic algorithm for resource-constrained project scheduling problems. In: International Conference on Intelligent Data Engineering and Automated Learning, pp. 202–209. Springer (2013)
27. Wang, L., Liu, J.: Solving multimode resource-constrained project scheduling problems using an organizational evolutionary algorithm. In: Proceedings of the 18th Asia Pacific Symposium on Intelligent and Evolutionary Systems, vol. 1, pp. 271–283. Springer (2015)
28. Wang, L., Liu, J., Zhou, M.: An organizational cooperative coevolutionary algorithm for multimode resource-constrained project scheduling problems. In: Asia-Pacific Conference on Simulated Evolution and Learning, pp. 680–690. Springer (2014)
29. Wright, S.: The roles of mutation, inbreeding, crossbreeding, and selection in evolution, vol. 1, na (1932)
30. Yuan, X., Xiao, C., Lv, X., Liu, J.: A multi-agent genetic algorithm for resource constrained project scheduling problems. In: Proceedings of the 15th Annual Conference Companion on Genetic and Evolutionary Computation, pp. 195–196. ACM (2013)

Chapter 4
Network-Based Problem Difficulty Prediction Measures

In the previous chapter, the analysis about problem difficulty from the viewpoint of networks has been introduced. However, the most important purpose of problem difficulty analysis is to design measures that can easily evaluate the difficulty of different types of problems for EAs. Many predictive measures have been proposed from different viewpoints, but very few were proposed from the viewpoint of complex networks. Therefore, in this chapter, we first briefly introduce existing predictive measures and then focus on measures designed from the viewpoint of complex networks.

4.1 Nonnetwork-Based Problem Difficulty Prediction Measures

The study of factors affecting the performance of EAs is one of the fundamental research fields within the EA community. During the last 20 years, many methods have been studied from different perspectives. In general, they can be divided into two classes [4]: (1) Methods belonging to the first class focus on the properties of a particular algorithm, such as the building block (BB) hypothesis [9], epistasis variance [6], and epistasis correlation [19], and (2) methods belonging to the second class focus on the problem itself and in particular on fitness landscapes, such as isolation [7], multi-modality [4], and fitness distance correlation [11].

In the first class, the BB hypothesis [9], which is famous in the genetic algorithm (GA) community, states that a GA tries to combine low-order and highly fit schemata. Following the BB hypothesis, Davidor and Naudts [6, 19] proposed the measures of epistasis variance and epistasis correlation to assess the GA hardness of problems from the perspective of theoretical genetics. In the context of optimization and evolutionary algorithms, epistasis denotes the interaction between the sites in

© Springer Nature Switzerland AG 2019
J. Liu et al., *Evolutionary Computation and Complex Networks*,
https://doi.org/10.1007/978-3-319-60000-0_4

the expression of the fitness function. However, none of these methods has fully succeeded in giving a reliable measure of GA hardness [4, 20].

While epistasis variance attempts to detect or measure the epistasis (amount of interaction between the sites) in a fitness function, it merely detects the absence of epistasis, which is also the case for epistasis correlation. Moreover, the epistasis variance measure suffers from a technical problem caused by the choice of reference class: It is easy to verify that the first-order functions contain all constant functions. Consequently, epistasis variance will return the value 0 both for constant fitness functions and other first-order functions. This is counterintuitive because EAs deal very differently with the two types of functions. The latter are optimized easily, while constant areas in the fitness landscape decrease the search efficiency. Another serious weakness of the epistasis measures is their sensitivity to nonlinear scaling, something a GA with ranking selection is entirely insensitive to. Thus, the epistasis measures cannot characterize easy functions; that is, neither high epistasis variance nor low epistasis correlation implies a difficult problem [20].

In the second class, the methods focus on using the statistical properties of fitness landscapes to characterize problem difficulty. Isolation (needle-in-a-haystack) [7] and multi-modality [4] were perhaps the first attempts to connect the fitness landscape to the complexity of a problem. However, existing studies showed that isolation might be sufficient but is not a necessary condition for a difficult landscape, and multi-modality is neither necessary nor sufficient for a landscape to be difficult to search [4, 12].

The other popular one is the fitness distance correlation [11], which measures the hardness of a landscape according to the correlation between the distance from the optimum and the fitness value of the solution. Despite good success, fitness distance correlation is not able to predict performance in some scenarios [1, 20]. First, fitness distance correlation is also sensitive to nonlinear scaling. Second, contrary to the epistasis measures, although fitness distance correlation can detect the presence of constantness in the landscapes and report a correlation of 0 for constant functions, given its construction based on averaging, fitness distance correlation can be blinded by the presence of a large proportion of irrelevant deception [20].

Recently, Lu et al. in [14] pointed out that evolvability is an important feature directly related to problem hardness for EAs, and a general relationship that holds for evolvability and problem hardness is the higher the degree of evolvability, the easier the problem is for EAs. They presented, for the first time, the concept of Fitness-Probability Cloud (*fpc*) to characterize evolvability from the point of view of escape probability and fitness correlation. Furthermore, a numerical measure called Accumulated Escape Probability (*aep*) based on *fpc* was also proposed to quantify this feature and therefore problem difficulty. To illustrate the effectiveness of the proposed approach, they applied it to four test problems: ONEMAX, Trap, OneMix, and Subset Sum. They then contrasted the predictions made by the *aep* to the actual performance measured using the number of fitness evaluations. The results suggested that the new measure can reliably indicate problem hardness for EAs.

In addition to fitness landscapes, Borenstein et al. [4] gave a new interpretation to the concept of "landscape." They pointed out that a limitation of the original fitness landscape approach is that it does not provide a way to quantify the amount of information available in a landscape nor to assess its quality. Thus, they proposed *Information Landscapes* based on tournament selection in GAs. Using information landscapes, they proposed a method to predict GA hardness and a theoretical model to study search algorithms [5, 23].

In addition to the study on proposing predictive measures, He et al. in [10] study problem difficulty itself. They gave a rigorous definition of difficulty measures in black-box optimization and proposed a classification. Motivated by the distinction between theoretical and empirical versions of statistical measures, they classified realizations of difficulty measure into two types: namely exact realizations and approximate realizations. For both types of realizations, it was proven that predictive versions that run in polynomial time in general do not exist unless certain complexity-theoretical assumptions are wrong.

Although none of the available predictive difficulty measures achieves full success, and He et al. [10] pointed out that no predictive difficulty measures can run in polynomial time, there are still some successful applications of using predictive difficulty measures to guide designing new algorithms. For example, Merz and Tavares et al. [16, 21] conducted a fitness landscape analysis for quadratic assignment problems and multi-dimensional knapsack problems, respectively, and designed algorithms accordingly. Liu et al. in [13] analyzed the effect of two representations, namely binary and permutation, on the difficulty of multi-dimensional knapsack problems using the proposed network-based predictive measures, namely motif difficulty. The results show that the permutation representation with a first-fit heuristic is better than binary representation.

4.2 Network-Based Problem Difficulty Prediction Measures

In this section, we introduce the problem difficulty prediction measures designed based on one important network property, namely network motifs [17]. Network motifs can be considered as simple building blocks of complex networks. It has been shown that network motifs exist widely in various complex networks from biochemistry, neurobiology, ecology, and engineering, and different complex networks have different types of motifs [17]. Thus, network motifs are in fact an intrinsic property of complex networks and can be used to differentiate different networks. For example, Ghoneim et al. [8] used network motifs to discriminate two-player strategy games and obtained good performances. For EAs, some fitness landscape networks are easy, and some are difficult. These networks themselves are expected to be different in some intrinsic properties. Network motifs, as the building blocks of complex networks, provide one such property.

Based on this, Liu et al. in [13] proposed a predictive difficulty measure for EAs, namely motif difficulty (MD), by extracting motif properties from directed fitness landscape networks. Distance motifs are first designed and then divided into three classes based on their contributions to the search process of EAs. Finally, the measure, MD, is designed by synthesizing the effect of different classes of motifs on the search process from two levels, namely network level and node level, according to the features of EAs. The experimental results show that MD is a good predictive difficulty measure for EAs and can quantify the difficulty of different problems into the range of -1.0 (easiest) to 1.0 (most difficult). MD performs especially well on some counterexamples for other difficulty measures, such as FDC and epistasis variance.

4.2.1 Motifs in Fitness Landscape Networks

Based on the definition of fitness landscapes, fitness landscape networks **FNL** are also defined by a tuple of three components:

$$\mathbf{FNL} = (\mathbf{V}, \mathbf{E}, \mathbf{W}) \tag{4.1}$$

where $\mathbf{V}$ is the set of nodes, and each node corresponds to one candidate solution in $\mathbf{S}$; $\mathbf{E}$ is the set of edges, and each edge connects one candidate solution to one of its neighbors. That is,

$$\mathbf{E} = \{(x, y) \mid x, y \in \mathbf{S} \text{ and } x \in \mathbf{N}(y)\} \tag{4.2}$$

$\mathbf{W}$ is the set of weights of all nodes, and each weight is equal to the fitness value of the corresponding candidate solution.

Complex networks are relevant for many fields of science. Many complex networks, either natural or man-made, show common global characteristics such as the small-world property (i.e., short paths between any two nodes), clustering (i.e., few highly connected nodes), and scale free (i.e., nodes' degree follows power law distribution). These characteristics describe the network as a whole [2].

Alternatively, for a deeper understanding of these complex networks, and going beyond their global features, "network motifs" were proposed in [17]. They are defined as patterns of interconnections occurring in complex networks at numbers that are significantly higher than those in randomized networks. In fact, network motifs are subgraphs of three to five nodes that appear more significantly in real networks than in randomly generated networks. That is, network motifs describe the local structures of the networks.

Milo et al. [17] proposed a procedure for detecting network motifs. They started with networks in which the interactions between nodes are represented by directed edges. Each network was scanned for all of the possible n-node subgraphs, and the number of occurrences of each subgraph was recorded. They showed that several networks (i.e., gene regulation, neurons, food webs, electronic circuits, and World Wide Web) exhibit different types of network motifs, and frequencies of different network motifs vary from one network to another. Motifs may thus define universal classes of networks and are the basic building blocks of most networks. Therefore, network motifs have been widely used in studying complex systems and in characterizing features on the system level by analyzing locally how the substructures are formed.

The execution process of an EA is actually navigating a fitness landscape network. That is to say, the evolutionary process of each individual is actually the process of visiting one node after another in the fitness landscape network. When moving from one node to another, it explores the local structure of the network. Thus, local structures determine whether or not the whole network is easy or not for an EA. As stated above, network motifs are a good characteristic to describe the local structures of networks.

The key problem in designing a predictive measure for a given EA is to find a characterization of a broad class of fitness functions on which this algorithm behaves similarly [20]. As is well known, one of the most important implications of detecting network motifs is grouping complex networks in superfamilies, assuming that similar factors affected their evolution and that these interconnected structures are performing similar tasks. Such an implication allows the usage of gained knowledge about one network to understand other networks in the same superfamily, which just accords with the objective of designing predictive measures for EAs. In fact, the broad class of fitness functions is equivalent to the superfamily of fitness landscape networks. Thus, the characterization of the broad class of fitness functions can be obtained by analyzing the interconnected structures that perform similar tasks in one superfamily of fitness landscape networks, and these can be realized by network motifs.

However, the original network motifs were proposed for directed graphs, whereas the fitness landscape networks are usually undirected ones. Therefore, to predict problem difficulty using network motifs, the fitness landscape networks need to be changed to directed graphs, which is accomplished by making use of the set of weights as follows.

Definition 4.1 Suppose a fitness landscape network to be **FNL** = (**V**, **E**, **W**), then the corresponding directed fitness landscape network is:

$$\overrightarrow{\textbf{FNL}} = (\textbf{V}, \overrightarrow{\textbf{E}}) \qquad (4.3)$$

where $\overrightarrow{\textbf{E}}$ is the set of directed edges, and each edge points from the node with lower weight to the node with higher weight. That is,

$$\overrightarrow{\textbf{E}} = \{\overrightarrow{(x, y)} \mid (x, y) \in \textbf{E} \text{ and } \textbf{W}(x) \leq \textbf{W}(y)\} \qquad (4.4)$$

where $\textbf{W}(.)$ stands for the weight of the node.

We know that the original network motifs are defined based on the difference in frequency between real networks and randomized networks [17]. But for EAs, a randomized fitness landscape network actually corresponds to a random problem, namely the problem with randomized fitness values for each candidate solution. Thus, randomized fitness landscape networks cannot be taken as the baseline to

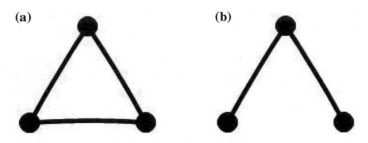

Fig. 4.1   Two kinds of motifs with three nodes in undirected graphs

design motifs, but the features of directed fitness landscape networks need to be analyzed.

In motif difficulty, only motifs with three nodes are considered. Clearly, without considering the directions of edges, there are only two kinds of motifs with three nodes: motifs with three nodes and three edges (Fig. 4.1a) and those with three nodes and two edges (Fig. 4.1b). Clearly, a motif with three nodes and three edges can be viewed as three motifs with three nodes and two edges. Thus, only the motifs with three nodes and two edges are considered.

Next, the directions of edges are considered. According to Definition 4.1, there are two edges with different directions connecting the same pair of nodes only when the weights of these two nodes are equal. Thus, for a pair of nodes (x, y), there are three types of connections, that is, $\overrightarrow{(x, y)}$ only, $\overrightarrow{(y, x)}$ only, and both $\overrightarrow{(x, y)}$ and $\overrightarrow{(y, x)}$. Based on these types of connections, at most, there are nine different types of motifs. However, some of them are equivalent. Finally, there are six types of different motifs, which are listed in Table 4.1 and labeled as **Group$_0$**–**Group$_5$**.

Because the search direction of EAs on motifs cannot be determined (i.e., we do not know whether the search would start at $\mathbf{n}_1$ or $\mathbf{n}_3$, and $\mathbf{n}_1$ and $\mathbf{n}_3$ are equivalent for EAs), the two types of motifs in **Group$_1$** are equivalent, and **Group$_2$** and **Group$_3$** are similar. Without loss of generality, in the following text, the first type of motifs in **Group$_1$**, and **Group$_2$** and **Group$_3$** are always used. What should be noted is we divided motifs into different groups without considering the connection type of $(\mathbf{n}_1, \mathbf{n}_3)$. This is because there is no edge connecting $\mathbf{n}_1$ and $\mathbf{n}_3$; namely, these two nodes cannot be transformed to each other in the evolutionary process. These six types of motifs can be viewed as the basic building blocks of a directed fitness landscape network.

Because selection pressure in EAs causes the search heuristic to prefer high-fitness regions, the problems would be easy if high-fitness regions are close to the global optimum. On the contrary, if high-fitness regions are far from the global optimum, selection pressure may mislead the search heuristic to a wrong direction. When the search process visits a motif from one end to the other end, the distance to the global optimum is changed. Thus, based on the distance of each node to the global optimum, **distance motifs** are defined as follows:

Table 4.1 Six types of motifs

No.	Motifs
Group$_0$	n_1 ⟷ n_2 ⟷ n_3
Group$_1$	n_1 ⟷ n_2 n_3 or n_1 n_2 ⟷ n_3
Group$_2$	n_1 ⟷ n_2 ⟷ n_3 or n_1 ⟷ n_2 → n_3
Group$_3$	n_1 ← n_2 → n_3 or n_1 → n_2 → n_3
Group$_4$	n_1 ⟷ n_2 → n_3
Group$_5$	n_1 → n_2 ⟷ n_3

Definition 4.2 A **distance motif** (**DM**) is a motif in which each node is associated with a distance, which is the distance from this node to the closest global optimum in phenotype space.

Here, the distance in phenotype space is used because the phenotype of a global optimum may be mapped into more than one genotype under some encoding methods. Clearly, distance motifs can be divided into different groups based on the distance of each node. Here, when the search process visits a motif from one end to the other end, the distances of the two ends should be compared to ensure the search process to head toward the right direction. Thus, the relationships about distances among three pairs are considered; that is, (n_1, n_2), (n_2, n_3), and (n_1, n_3). For the sake of simplicity, only two relationships are considered, namely $\leq$ and $\geq$. As a result, the six types of motifs can be further divided into 36 groups at most. When taking into account the effect of fitness values, 21 different groups are determined, which are shown in Table 4.2, where d_1, d_2, and d_3 denotes the distance of n_1, n_2, and n_3, respectively.

In fact, these 21 groups are obtained by dividing **Group$_0$** to **Group$_5$** into subgroups and are numbered as **Group$_0$**, **Group$_{1,1}$**–**Group$_{1,4}$**, **Group$_{2,1}$**–**Group$_{2,4}$**, **Group$_{3,1}$**–**Group$_{3,6}$**, **Group$_{4,1}$**–**Group$_{4,3}$**, and **Group$_{5,1}$**–**Group$_{5,3}$**. For **Group$_0$**, since all fitness values are equal, no subgroups are generated. For **Group$_1$**, since $w_1 = w_2$, no subgroups are generated based on the relationship between d_1 and d_2, and the sub-

Table 4.2 Twenty-one groups of different distance motifs

Motifs	No.	Distance Motifs
n_1 ⬤⬤ n_2 ⬤ n_3	**Group$_0$**	All cases
n_1 ⬤ n_2 ⬤ n_3 ⬤	**Group$_{1,1}$**	$d_1 \geq d_2 \geq d_3$ or $d_2 \geq d_1 \geq d_3$
	Group$_{1,2}$	$d_3 \geq d_1 \geq d_2$ or $d_3 \geq d_2 \geq d_1$
	Group$_{1,3}$	$d_1 \geq d_3 \geq d_2$
	Group$_{1,4}$	$d_2 \geq d_3 \geq d_1$
n_1 ⬤ n_2 ⬤ n_3 ⬤	**Group$_{2,1}$**	$d_3 \geq d_2 \geq d_1$ or $d_2 \geq d_3 \geq d_1$
	Group$_{2,2}$	$d_1 \geq d_3 \geq d_2$ or $d_1 \geq d_2 \geq d_3$
	Group$_{2,3}$	$d_3 \geq d_1 \geq d_2$
	Group$_{2,4}$	$d_2 \geq d_1 \geq d_3$
n_1 ⬤ n_2 ⬤ n_3 ⬤	**Group$_{3,1}$**	$d_1 \geq d_2 \geq d_3$
	Group$_{3,2}$	$d_1 \geq d_3 \geq d_2$
	Group$_{3,3}$	$d_2 \geq d_1 \geq d_3$
	Group$_{3,4}$	$d_2 \geq d_3 \geq d_1$
	Group$_{3,5}$	$d_3 \geq d_1 \geq d_2$
	Group$_{3,6}$	$d_3 \geq d_2 \geq d_1$
n_1 ⬤ n_2 ⬤ n_3 ⬤	**Group$_{4,1}$**	$d_1 \geq d_3 \geq d_2$ or $d_3 \geq d_1 \geq d_2$
	Group$_{4,2}$	$d_2 \geq d_1 \geq d_3$ or $d_2 \geq d_3 \geq d_1$
	Group$_{4,3}$	$d_1 \geq d_2 \geq d_3$ or $d_3 \geq d_2 \geq d_1$
n_1 ⬤ n_2 ⬤ n_3 ⬤	**Group$_{5,1}$**	$d_1 \geq d_3 \geq d_2$ or $d_3 \geq d_1 \geq d_2$
	Group$_{5,2}$	$d_2 \geq d_1 \geq d_3$ or $d_2 \geq d_3 \geq d_1$
	Group$_{5,3}$	$d_1 \geq d_2 \geq d_3$ or $d_3 \geq d_2 \geq d_1$

groups are obtained by considering the relationship between d_3 and d_1, d_2. When d_3 is in the middle, the relationship between (d_2, d_3) is different, and since w_2 and w_3 are different, they are divided into two subgroups. **Group**$_2$ is similar to **Group**$_1$, just considering the position of d_1 instead. For **Group**$_3$, since all weights are different, the relationships between each pair are considered, with six subgroups being generated. For **Group**$_4$, since the relationship between w_1 and w_3 is unknown, the subgroups are obtained by considering only the position of d_2. **Group**$_5$ is similar.

The search process of EAs actually visits a sequence of different kinds of distance motifs, and the results can also be seen as the accumulated effect of these motifs. Thus, to predict problem difficulty, we can start from a basic level; that is, analyze the effect of each kind of motifs on the search process, and distinguish which kinds of motifs can help the search process to head toward the right direction and which kinds of motifs prevent the search process from heading toward the right direction.

Next, based on the characteristics of the search process of EAs, the effect of various distance motifs on problem difficulty is analyzed from two perspectives. The first is from the motifs themselves. Since motifs are basic building blocks of the whole networks, the statistics on motifs reflect some global feature of the network. The difficulty measure designed from this perspective is named as **network-level difficulty**. The second is based on the characteristics of the search process, that is, selection pressure in EAs. How each node is involved in various kinds of motifs and their effects on problem difficulty are analyzed. Thus, the difficulty measure designed from this perspective is named as **node-level difficulty**. Finally, network and node-level difficulties are integrated to define a measure, **motif difficulty**.

4.2.2 Network-Level Prediction Measure

As stated above, the performance of EAs can be viewed as the accumulated effect of the motifs they have visited. Thus, each motif EAs have visited has contributions on the final performance. Based on such contributions, three classes of motifs are defined as follows:

Definition 4.3 A **guide motif** is a distance motif, for which the distance decreases when the nodes' weights increase. A **deceptive motif** is a distance motif, for which the distance increases when the nodes' weights increase. A **neutral motif** is a distance motif for which all nodes have identical weights.

In guide motifs, when the search process goes from low-fitness to high-fitness nodes, it approaches the global optimum. Thus, guide motifs guide the search process in the right direction. On the contrary, in deceptive motifs, when the search process goes from low-fitness to high-fitness nodes, it is apart from the global optimum. Thus, deceptive motifs are deceptions for EAs and increase problem difficulty. Neutral motifs actually correspond to neutral landscapes. They do not increase problem difficulty, but also do not decrease it.

Table 4.3 Class of each kind of distance motif

Guide motifs		Neutral motifs	Deceptive motifs	
Group$_{1,2}$	**Group**$_{3,5}$	**Group**$_0$	**Group**$_{1,1}$	**Group**$_{3,3}$
Group$_{1,3}$	**Group**$_{3,6}$		**Group**$_{1,4}$	**Group**$_{4,1}$
Group$_{2,1}$	**Group**$_{4,2}$		**Group**$_{2,2}$	**Group**$_{4,3}$
Group$_{2,4}$	**Group**$_{5,1}$		**Group**$_{2,3}$	**Group**$_{5,2}$
Group$_{3,4}$			**Group**$_{3,1}$	**Group**$_{5,3}$
			Group$_{3,2}$	

Clearly, if a fitness landscape network is composed of guide motifs, the corresponding problem is easy. Because selection pressure in EAs makes the search process approach higher fitness solutions ceaselessly, and the guide motifs guarantee each time the search process reaches a higher fitness solution, it gets closer to the global optimum. Therefore, the most straightforward way to predict problem difficulty by motifs is to calculate the percentages of guide motifs, deceptive motifs, and neutral motifs. Thus, we need to determine, among the 21 groups of distance motifs (see Table 4.2), which groups belong to guide motifs, which belong to deceptive motifs, and which belong to neutral motifs.

Based on the definitions of the three classes of motifs, **Group**$_0$ belongs to neutral motifs; **Group**$_{1,2}$, **Group**$_{2,1}$, and **Group**$_{3,6}$ are guide motifs; and **Group**$_{1,1}$, **Group**$_{2,2}$, and **Group**$_{3,1}$ are deceptive motifs. For **Group**$_{1,3}$ and **Group**$_{1,4}$, since $w_1 = w_2$, their types are determined by the relationship between n_2 and n_3; namely, **Group**$_{1,3}$ belongs to guide motifs, and **Group**$_{1,4}$ belongs to deceptive motifs. **Group**$_{2,3}$ and **Group**$_{2,4}$ are similar, and they belong to deceptive and guide motifs, respectively. **Group**$_{3,2}$ to **Group**$_{3,5}$ do not accord with the definitions strictly, and we consider only situations of two end nodes. Thus, **Group**$_{3,2}$ and **Group**$_{3,3}$ are categorized as deceptive motifs, and **Group**$_{3,4}$ and **Group**$_{3,5}$ are categorized as guide motifs. For **Group**$_4$, we use the situation of n_2 to distinguish their types. Thus, **Group**$_{4,1}$ is deceptive motifs, **Group**$_{4,2}$ is guide motifs, and **Group**$_{4,3}$ is categorized as deceptive motifs. **Group**$_5$ is similar, **Group**$_{5,1}$ is guide motifs, **Group**$_{5,2}$ is deceptive motifs, and **Group**$_{5,3}$ is also categorized as deceptive motifs. The class of each group is summarized in Table 4.3.

Based on these three classes of motifs, the **network-level difficulty** is defined as follows:

Definition 4.4 For a directed fitness landscape network, $\overrightarrow{\textbf{FLN}}$, $|\textbf{DiM}_{\overrightarrow{\textbf{FLN}}}|$ is the number of all distance motifs in $\overrightarrow{\textbf{FLN}}$, $|\textbf{GM}_{\overrightarrow{\textbf{FLN}}}|$ and $|\textbf{DeM}_{\overrightarrow{\textbf{FLN}}}|$ are the numbers of guide motifs and deceptive motifs in $\overrightarrow{\textbf{FLN}}$, respectively. Then, **network-level difficulty**, *NeLD*, is defined as follows:

$$NeLD = \frac{|\textbf{DeM}_{\overrightarrow{\textbf{FLN}}}| - |\textbf{GM}_{\overrightarrow{\textbf{FLN}}}|}{|\textbf{DiM}_{\overrightarrow{\textbf{FLN}}}|} \tag{4.5}$$

The range of network-level difficulty is $[-1.0, 1.0]$. When *NeLD* is equal to -1.0, all distance motifs are guide motifs, so the problem is the easiest. On the contrary, when *NeLD* is equal to 1.0, all are deceptive motifs, so the problem is the most difficult. This definition accords with the intuitions; namely, the more deceptive motifs exist, the more difficult a problem is. In fact, this is also true for some problems. For example, we all know *ONEMAX* problems are easy and *RIDGE* problems [15] are difficult for EAs. The following experimental results show that all distance motifs in *ONEMAX* problems' fitness landscape networks are guide motifs and that most distance motifs in RIDGE problems' fitness landscape network are deceptive motifs. Therefore, network-level difficulty has a predictive power for problem difficulty.

4.2.3 Node-Level Prediction Measure

In network-level difficulty, only the percentage of each class of motifs is considered. When the percentage of some class of motifs is high enough, their effect on problem difficulty is clear. However, when the percentage is not high enough, even if they exist, they may have no effect on problem difficulty. Thus, the difficulty measure based only on the percentage of each class of motifs may be ineffective in some situations. The following experiments on *DECEPTIVE_MIXTURE* problems [20] also confirm this. The most important characteristic of the search process of EAs is that it works under selection pressure. Some trajectories are preferred over others, which make it impossible to visit some paths during the search process. That is to say, although the motifs on such paths exist, they have no effect on problem difficulty. So, we need to take into account the effect of selection pressure when designing difficulty measures.

How selection pressure works is reflected by what kinds of paths are chosen; namely, how the search process moves from one node to another. From the viewpoint of nodes, each node is a part of many motifs, and these motifs may belong to different classes. Under selection pressure, guide motifs are always preferred. Even if a node is a part of both guide motifs and deceptive motifs, the search process may not visit the deceptive motifs at all. Therefore, we propose another difficulty measure, named as node-level difficulty. To define this measure, we need first to distinguish different types of nodes. Here, we classify nodes based on its associated set of motifs.

Among the 21 groups of distance motifs, **Group**$_{1,2}$, **Group**$_{2,1}$, and **Group**$_{3,6}$ strictly accord with the definition of guide motifs, while **Group**$_{1,1}$, **Group**$_{2,2}$, and **Group**$_{3,1}$ strictly accord with the definition of deceptive motifs. Moreover, their effect on the search process is also explicit; thus, the nodes in these motifs are explicitly affected by selection pressure. Therefore, two types of nodes are defined, namely **Core Guide Nodes** and **core deceptive nodes**, as follows:

Definition 4.5 For a node **n**, if one of the distance motifs that **n** is a part of belongs to **Group**$_{1,2}$, **Group**$_{2,1}$, or **Group**$_{3,6}$, and **n** is the third node, namely n_3 in this motif, then **n** is a **core guide node**. If one of the distance motifs that **n** is a part of belongs to $\mathbf{Group}_{1,1}$, $\mathbf{Group}_{2,2}$, or $\mathbf{Group}_{3,1}$, and $\mathbf{n}$ is the first node, namely $\mathbf{n}_1$ in this motif, then $\mathbf{n}$ is a **core deceptive node**.

Intuitively, any node that is a part of guide motifs is helpful for leading the search process to a right direction, and any node that is a part of deceptive motifs is harmful. However, after we analyze motifs in detail, we find that the three nodes play different roles. Specifically, in the case of motifs that belong to $\mathbf{Group}_{3,6}$, only the third node is a core guide node. Because when the search process visits this group of motifs from $\mathbf{n}_3$ to $\mathbf{n}_1$, it is heading toward the right direction. But when the search process reaches $\mathbf{n}_1$, all other motifs that $\mathbf{n}_1$ is a part of may be deceptive ones, and the search process cannot continuously head toward the right direction. Thus, we cannot determine whether the contribution of $\mathbf{n}_1$ is positive or negative. For similar reasons, only the first nodes of motifs belonging to $\mathbf{Group}_{3,1}$ are core deceptive nodes.

On the other hand, the effect of the nodes in $\mathbf{Group}_{3,1}$ and $\mathbf{Group}_{3,6}$ is clear, and one may argue the same for the effect of the nodes in $\mathbf{Group}_{1,1}$, $\mathbf{Group}_{1,2}$, $\mathbf{Group}_{2,1}$, and $\mathbf{Group}_{2,2}$. In fact, these four groups represent the boundary of the neutral landscapes. Thus, such nodes can lead the search process to exit or join neutral landscapes.

Based on the core guide and deceptive nodes, the **node-level difficulty** is defined as follows:

Definition 4.6 For a directed fitness landscape network, $\overrightarrow{\mathbf{FLN}} = (\mathbf{V}, \overrightarrow{\mathbf{E}})$, $\mathbf{V}_{CGN}$ is the set of core guide nodes, and $\mathbf{V}_{CDN}$ is the set of core deceptive nodes. Then, **node-level difficulty**, *NoLD*, is defined as follows:

$$NoLD = \frac{|(\mathbf{V} - \mathbf{V}_{CGN}) \cap \mathbf{V}_{CDN}| - \mathbf{V}_{CGN}}{\mathbf{V}} \tag{4.6}$$

According to Definition 4.5, any node can be a core guide node and core deceptive node simultaneously. To embody the influence of selection pressure, node-level difficulty is defined by preferring the core guide nodes. That is, once a node is a core guide node, its effect is taken into account, and only when a node is a core deceptive node but not a core guide node, its effect can be taken into account. Thus, $(\mathbf{V} - \mathbf{V}_{CGN}) \cap \mathbf{V}_{CDN}$ denotes the set of core deceptive nodes that are not core guide nodes. The range of node-level difficulty is still $[-1.0, 1.0]$, and -1.0 corresponds to the easiest problems, while 1.0 corresponds to the most difficult problems.

4.2.4 Measure Combining Network and Node-Level Prediction Measures

Both network-level difficulty and node-level difficulty are based on distance motifs, and they describe problem difficulty from different angles. Node-level difficulty mainly reflects the effect of selection pressure. Since selection pressure has the dominant effect in EAs, node-level difficulty can predict the general problem difficulty.

However, node-level difficulty reflects only the number of nodes that are involved in guide or deceptive motifs, but not how the nodes are involved in these motifs, namely the number of guide or deceptive motifs they are actually a part of. Thus, node-level difficulty cannot reflect the detailed difference on problem difficulty. The following experiments on *DECEPTIVE_MIXTURE* problems also confirm this. Clearly, this detailed difference can be reflected by network-level difficulty. Therefore, the above two measures are integrated together to form a new measure, namely **motif difficulty**, to reflect both the general problem difficulty and the detailed differences.

Definition 4.7 For a directed fitness landscape network, $\overrightarrow{FLN}$, **motif difficulty** (*MD*), is defined as follows:

$$MD = \frac{NeLD + NoLD}{2} \tag{4.7}$$

The range of node-level difficulty is still $[-1.0, 1.0]$, and -1.0 corresponds to the easiest problems while 1.0 corresponds to the most difficult problems. In fact, node-level difficulty is equivalent to predicting problem difficulty from a macro-level, while network-level difficulty is from a microlevel. Motif difficulty just integrates the predicting results of these two levels together.

4.2.5 Performance of Network-Based Prediction Measures

If a measure is computed on the whole search space, its values are said to be exact [20]. Thus, all experiments in this section computed the value of motif difficulty exhaustively on the whole search space. We first investigate MD's prediction of EAs' behavior on a number of reasonably well-studied fitness functions. A collection of fitness functions of known difficulty and different characteristics is used to validate the MD's prediction under various situations. Then, MD is used to estimate the difficulty of three counterexamples for other difficulty measures.

In the following experiments, a candidate solution in search spaces is a binary string, $\mathbf{s} = (s_1, s_2, \ldots, s_n)$, where s_i, $i = 1, 2, \ldots, n$ is "0" or "1". '*' stands for a wildcard symbol, and $s^{(m)} = ss\ldots s$, where s can be 0, 1, or *. Moreover, #1(**s**) means the number of "1" in the string **s**. The length of binary strings for all fitness functions is set to 16. The fitness landscape networks are built using a one-bit flip operator. The distance to global optima is given by the Hamming distance.

To validate MD's predictions, the same GA is used to observe the hardness of fitness functions. This GA is equipped with binary tournament selection, uniform crossover, and $1/n$ mutation rate. The population size is 100, and the maximum number of generations is set to 1,000. The first generation in which the global optimum was found is used as the performance measure. If the global optimum is not found within the maximum number of generations, then the performance is set to the maximum number of generations. To reduce the effect of randomization, all results are averaged over 1,000 runs with different randomly sampled initial populations.

4.2.5.1 Experiments on well-studied fitness functions

The fitness functions used in this subsection are summarized in Table 4.4. These fitness functions have different characteristics, which result in different types of fitness landscapes. *ONEMAX* and *LEADING_ONES* [15] functions have reliable information, while *RIDGE* [15] and *LV_DECEPTIVE* [22] functions have unreliable information. The fitness landscape of *NIAH* function is flat while that of *RAND* function is random, and these two functions have neither reliable information nor unreliable information. Furthermore, fitness functions with a variable level of difficulty are also studied, that is, $MULTI\_MODAL_i$ function [3] with a varying number of local maximum i and $TRAP_i$ $(i = 1, 2, \ldots, n)$ [3] whose difficulty is controlled by i. The diversity of these fitness landscapes is useful to validate MD's performance under various situations.

Table 4.4 Well-studied fitness functions

Fitness function names	Definitions
ONEMAX	$ONEMAX(s) = \sum_{i=1}^{n} s_i$
LEADING_ONES [15]	$LEADING\_ONES(s) = \sum_{i=1}^{n} \prod_{j=1}^{i} s_j$
NIAH	The global optimum is first selected uniformly at random from the search space, and then the fitness value of the global optimum is set to 1 while those of all other candidate solutions are set to 0
RAND	The global optimum is first selected uniformly at random from the search space, and then the fitness value of the global optimum is set to 2^n while those of all other candidate solutions are chosen uniformly at random from $[1, 2^n - 1]$
RIDGE [15]	$RIDGE(s) = \begin{cases} n + 1 + \#1(s) & \text{if } \exists i \in \{0, 1, \ldots, n\}, s.t.\ s = 1^{(i)}0^{(n-i)} \\ n - 1\#(s) & \text{otherwise} \end{cases}$
LV_DECEPTIVE [22]	$LV\_DECEPTIVE(s) = \begin{cases} 2n & \text{if } \#1(s) = n \\ 2n - 1 & \text{if } \#1(s) = 0 \\ 2n - 2(\#1(s)) - 2 & \text{otherwise} \end{cases}$
$MULTI\_MODAL_i$ [3]	The global optimum is first selected uniformly at random from the search space, and its fitness value is set to $2n$. Then, i different local optima are selected uniformly at random from the search space, and their fitness values are set to n. Finally, the fitness values of all other candidate solutions are computed as n minus the distance to the closest local or global optimum
$TRAP_i$ [3]	$TRAP_i(s) = \begin{cases} 2n & \text{if } \#1(s) = 0 \\ n - \#1(s) & \text{else if } \#1(s) < i \\ \#1(s) & \text{else if } \#1(s) \geq i \end{cases}$

Table 4.5 Experimental results on the former six fitness functions

Fitness function names	MD	GA performance (#generations)
ONEMAX	−0.9999	6.212
LEADING_ONES	−0.6167	20.095
NIAH	−0.0009	519.761
RAND	−0.1158	515.802
RIDGE	0.9971	999.999
LV_DECEPTIVE	0.9989	999.999

The values of MD and GA performance for the former six fitness functions are given in Table 4.5. The GA performance shows that *ONEMAX* is extremely easy, and *LEADING_ONES* is more difficult than *ONEMAX*, but is still easy since the average number of generations is only 20.095. *NIAH* and *RAND* have similar difficulty, whose average number of generations is both around 500. *RIDGE* and *LV_DECEPTIVE* are extremely difficult since GA nearly cannot find the global optimum in all of the 1,000 runs. These are also the available knowledge on these problems, and the values of MD listed in Table 4.5 clearly confirm this knowledge. The value of MD for *ONEMAX* is −0.9999, which is nearly the lowest and namely the easiest problem. The value of MD for *LEADING_ONES* is −0.6167, which illustrates that it is harder than *ONEMAX*, but still belongs to easy problems since the value of MD is smaller than −0.5. The values of MD for *NIAH* and *RAND* are similar, and both are around 0, which shows they belong to neutral landscapes. The values of MD for *RIDGE* and *LV_DECEPTIVE* are nearly equal to each other (0.9971 and 0.9989) and approach the highest value, indicating the most difficult types of problems.

$MULTI\_MODAL_i$ function is selected to test how MD reflects the effect of multi-modality on problem difficulty. The number of local optima varies from 1 to 30 to see the changes of the value of MD. The experimental results are given in Fig. 4.2, which shows that multi-modality is not a good indicator of problem difficulty. There is no clear decreasing or increasing trend on MD with the number of local optima. When the number of local optima increases to 10, it seems that problem difficulty just fluctuates in the range of −0.5 to −0.3. It illustrates that the common intuition that the more local optima a problem has, the more difficult it is, is incorrect. A function with three local maxima has the same expected difficulty as functions with 9 or 20 local maxima, and a function with three local maxima is more difficult than a function with 18 local optima. The GA performance shows the same phenomenon. Therefore, MD confirms the previous results regarding GA hardness; that is, multi-modality is neither necessary nor sufficient for a landscape to be difficult to search [12].

The $TRAP_i$ function is selected to test whether MD can reflect the changes of problem difficulty step by step. Clearly, the global optimum of $TRAP_i$ is the string of all 0, and $TRAP_i$ is defined as *ZEROMAX* for all candidate solutions with a distance to the global optimum smaller than i and *ONEMAX* for all other candidate solutions.

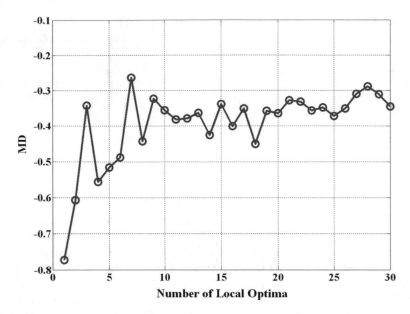

Fig. 4.2 $MULTI\_MODAL_i$ function

When i varies from 1 to n, problem difficulty decreases gradually. In fact, this reflects the effect of isolation on problem difficulty; that is, the more the global optimum is isolated from the other high-fitness solutions, the more difficult the problem is. The experimental results are shown in Fig. 4.3, which clearly illustrates that MD really reflects the gradual changes of problem difficulty. When i increases from 1 to n (16), the value of MD monotonically nonincreases from 0.9989 to -0.9998. Therefore, MD not only confirms the previous results about the effect of isolation on GA hardness but also shows the ability in reflecting gradual changes of problem difficulty.

The above experimental results show that MD confirms previous results on problem difficulty, including isolation, deception, multi-modality, neutrality, and randomness. Thus, MD is a meaningful indicator of problem difficulty.

4.2.5.2   Experiments on Counterexamples for Other Difficulty Measures

To further test the performance of MD, three classes of problems (scaling problems, constantness problems, and irrelevant deception problems) are used in this experiment. In fact, it has been shown that other difficulty measures fail to predict the difficulty of these three classes of problems. Epistasis measures, including epistasis variance [4] and epistasis correlation [19], cannot correctly predict the difficulty of constantness and scaling problems [20]. Although FDC can detect the presence

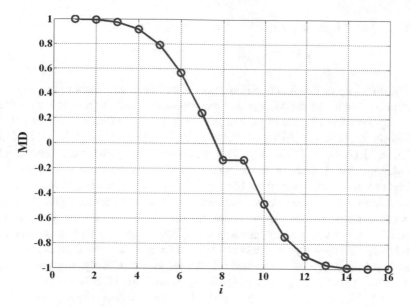

Fig. 4.3 Experimental results on $TRAP_i$ function

of constantness in the landscapes, FDC can be blinded by the presence of a large proportion of irrelevant deception since its construction is based on averaging [20].

Naudts et al. [20] constructed a scaling problem, namely *NML*, as follows:

$$NML_m(s) = \left(\frac{\#1(s)}{n} \right)^m \tag{4.8}$$

Clearly, the higher the control parameter m is, the more the fitness value of the global optimum; namely, the string of all "1" gets isolated from the mass of fitness values, e.g., a fitness distance correlation going to 0 for both m and n going to infinity [20]. In fact, this class of functions is a modified version of the *ONEMAX* function. For a GA with comparison-based selection, the difficulty of this class of functions does not change with the control parameter m. In the experiments, the GA performance is always about six generations for $m = 1, 2, \ldots, 30$ which is similar to that for *ONEMAX*. The value of MD is -0.9999 for any m, which is consistent with the performance of GA and is also equal to that of *ONEMAX*. Thus, it illustrates that MD has the advantage of being insensitive to nonlinear scaling.

Naudts et al. [20] constructed a generalized ROYAL ROAD function [18] as follows:

$$GENERALIZED\_ROYAL\_ROAD_{m,m'}(s) = \sum_{k=m}^{m'} 2k \left| \left\{ i \mid s \in H_i^{\log n,k} \right\} \right| \tag{4.9}$$

where $0 \leq m \leq log n$, $0 \leq i \leq 2^{log n - m}$, $m \leq m' \leq log n$, and

$$H_i^{log \ n, m} = *^{(2^{log \ m} i)} 1^{(2^{log \ m})} *^{(2^{log \ n} - 2^m (i+1))} \qquad (4.10)$$

The difficulty of this function is controlled by m and m'. Here, since $log \ n = 4$, experiments on all 15 pairs of (m, m') are conducted, and the results are shown in Fig. 4.4, where the 15 pairs of parameters are ordered from easy to difficult (1–15) as follows on the X-axis: $(0, 0)$, $(0, 1)$, $(0, 2)$, $(0, 3)$, $(0, 4)$, $(1, 1)$, $(1, 2)$, $(1, 3)$, $(1, 4)$, $(2, 2)$, $(2, 3)$, $(2, 4)$, $(3, 3)$, $(3, 4)$, $(4, 4)$. Figure 4.4a shows both the changes of NeLD and MD, and Fig. 4.4b shows the changes of the GA performance as a reference.

Figure 4.4a shows that both NeLD and MD reflect the difficulty changes in this class of functions. For the former five functions, the performances of NeLD and MD are similar, but for the later nine functions, the values of NeLD are always higher than those of MD. The GA performance shows that the functions with former 12 pairs of parameters are relatively easy. MD is -0.8749 for $m = 1$ and -0.6025 for $m = 2$, while NeLD is -0.75 for $m = 1$ and -0.225 for $m = 2$. Clearly, when $m = 1$ or 2, MD is always smaller than -0.5, which is more consistent with the GA performance than NeLD. For $m = 3$, the results are similar. The function with $(4, 4)$ is actually equivalent to a *NIAH* function, so both the difficulty measures and GA performance are similar to those for *NIAH* function.

This experiment illustrates that MD is more effective than NeLD, and NeLD alone cannot predict problem difficulty accurately. Moreover, for functions with similar difficulty, the value of MD does not change. For example, for $m = 0, 1, 2, 3, 4$, the value of MD is constant for each level. Thus, MD can detect the presence of constantness in the landscapes.

Naudts et al. [20] constructed a problem with irrelevant deception, namely *DECEPTIVE_MIXTURE*, by mixing an ONEMAX and a number of ZEROMAX problems as follows:

$$DECEPTIVE\_MIXTURE_m(s) =$$

$$\begin{cases} n - 1 - \#0(s_2, s_3, ...s_n) & \text{if } s_1 = 0 \\ n - 2 - \#0(s_3, s_4, ...s_n) & \text{else if } s_1 = 1 \text{ and } s_2 = 0 \\ ... & ... \\ n - m - \#0(s_{m+1}, s_{m+2}, ...s_n) + 2(m-1)n & \text{else if } s_1 = s_2 = ... = s_{m-1} = 1 \text{ and } s_m = 0 \\ \#1(s_{m+1}, s_{m+2}, ...sn) + 2mn & \text{else if } s_1 = s_2 = ... = s_m = 1 \end{cases}$$

$$(4.11)$$

Both fitness distance correlation and the sitewise optimization measure (a generalization for FDC and epistasis) [20] failed to correctly predict the performance for this function. The result in [3] indicates that the difficulty of this function is 0.75. However, the real difficulty of this function is not as high as that. On the other hand, for this function, the higher the mixture coefficient m is, the harder the problem is. But no available predictive measure is able to predict this.

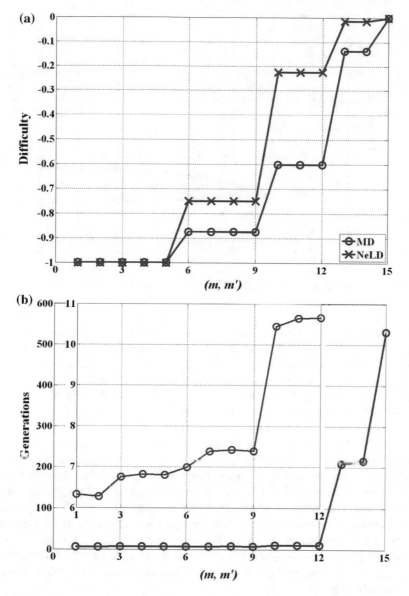

Fig. 4.4 Experimental results on *GENERALIZED_ROYAL_ROAD* function. **a** Predictive difficulties and **b** GA performance

Here, experiments on $m = 1$ (easy), $2, \ldots 8$ (difficult) are conducted, and the results are shown in Fig. 4.5, where the results of MD, NeLD, and FDC are all presented. First, the value of FDC decreases with m, which is in contradiction with the characteristics of this class of functions. For NeLD and MD, although both of

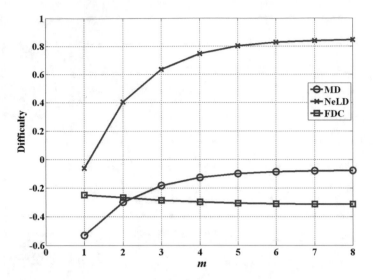

Fig. 4.5 Experimental results on *DECEPTIVE_MIXTURE* function

their values increase with m, NeLD is much larger than MD. The value of NeLD varies from -0.0625 to 0.8454. Most of them are greater than 0, which predicts that this class of functions is difficult. However, the analysis in [20] shows that EAs can solve this class of functions easily, and the deceptive information in this function is irrelevant to the search process of EAs. The values of MD vary from -0.5311 to -0.0772, which are smaller than 0 and predict this class of functions is not very difficult for EAs. Thus, MD is more accurate than NeLD.

Similar to the experiment in the last subsection, this experiment also illustrates that MD is more effective than NeLD. Moreover, the predicting results of MD are completely consistent with the characteristics of this class of functions; that is, the higher the mixture coefficient m is, the harder the problem is. Thus, MD can distinguish irrelevant information existing in the landscapes.

The above experimental results show that MD performs very well on the three classes of counterexamples for other difficulty measures. MD is insensitive to nonlinear scaling, which is consistent with GA, can detect the presence of constantness, and is robust to irrelevant deceptive information. Thus, MD is a good indicator of problem difficulty.

Since network motifs are basic building blocks of complex networks, problem difficulty is expected to be related to these building blocks, and the above work confirmed this. In fact, network motifs can be seen as the features of problems, and problem difficulty is determined by these features. Thus, to predict problem difficulty is equivalent to learning the relationship between these features and problem difficulty, which is actually a machine learning problem. This study represents an attempt toward finding such relationships in a simple way.

More work remains to be done. First, taking global optimum as reference is impractical for practical applications; thus, other reference points need to be investigated. Second, to study problem difficulty is only a step toward designing more efficient algorithms. MD will be useful in designing efficient search techniques or operators.

References

1. Altenberg, L., et al.: Fitness distance correlation analysis: an instructive counterexample. In: ICGA, pp. 57–64 (1997)
2. Barabási, A.L., Albert, R.: Emergence of scaling in random networks (1999)
3. Borenstein, Y.: Problem hardness for randomized search heuristics with comparison-based selection: a focus on evolutionary algorithms. Ph.D. thesis, University of Essex, UK (2008)
4. Borenstein, Y., Poli, R.: Information landscapes. In: Proceedings of the 7th Annual Conference on Genetic and Evolutionary Computation, pp. 1515–1522. ACM (2005)
5. Borenstein, Y., Poli, R.: Information landscapes and problem hardness. In: Proceedings of the 7th Annual Conference on Genetic and Evolutionary Computation, pp. 1425–1431. ACM (2005)
6. Davidor, Y.: Epistasis Variance: A Viewpoint on GA-Hardness, pp. 23–35 (1991)
7. Forrest, S., Mitchell, M.: Relative building block fitness and the building block hypothesis. In: D. Whitley (ed.) Proceedings of Foundations of Genetic Algorithms, pp. 109–126. Morgan Kaufmann, San Mateo, CA (1993)
8. Ghoneim, A., Abbass, H., Barlow, M.: Characterizing game dynamics in two-player strategy games using network motifs. IEEE Trans. Syst. Man Cybern. Part B (Cybern.) 38(3), 682–690 (2008)
9. Goldberg, D.E.: Genetic Algorithms in Search, Optimization, and Machine Learning. Addison-Wesley, Reading (1989)
10. He, J., Reeves, C., Witt, C., Yao, X.: A note on problem difficulty measures in black-box optimization: classification, realizations and predictability. Evol. Comput. 15(4), 435–443 (2007)
11. Jones, T., Forrest, S.: Fitness distance correlation as a measure of problem difficulty for genetic algorithms. In: Eshelman, L.J. (ed.) 6th International Conference Genetic Algorithms, pp. 184–192. Morgan Kaufmann, San Mateo, CA (1995)
12. Kallel, L., Naudts, B., Reeves, C.R.: Properties of fitness functions and search landscapes. In: Theoretical Aspects of Evolutionary Computing, pp. 175–206. Springer (2001)
13. Liu, J., Abbass, H.A., Green, D.G., Zhong, W.: Motif difficulty (md): a predictive measure of problem difficulty for evolutionary algorithms using network motifs. Evol. Comput. 20(3), 321–347 (2012)
14. Lu, G., Li, J., Yao, X.: Fitness-probability cloud and a measure of problem hardness for evolutionary algorithms. In: European Conference on Evolutionary Computation in Combinatorial Optimization, pp. 108–117. Springer (2011)
15. Malan, K.M., Engelbrecht, A.P.: Fitness landscape analysis for metaheuristic performance prediction. In: Recent Advances in the Theory and Application of Fitness Landscapes, pp. 103–132. Springer (2014)
16. Merz, P., Freisleben, B.: Fitness landscape analysis and memetic algorithms for the quadratic assignment problem. IEEE Trans. Evol. Comput. 4(4), 337–352 (2000)
17. Milo, R., Shen-Orr, S., Itzkovitz, S., Kashtan, N., Chklovskii, D., Alon, U.: Network motifs: simple building blocks of complex networks. Science 298(5594), 824–827 (2002)
18. Mitchell, M., Forrest, S., Holland, J.H.: The royal road for genetic algorithms: fitness landscapes and GA performance. In: Proceedings of the First European Conference on Artificial Life, pp. 245–254 (1992)

19. Naudts, B.: Measuring GA-hardness. Ph.D. thesis, University of Antwerp, Belgium (1998)
20. Naudts, B., Kallel, L.: A comparison of predictive measures of problem difficulty in evolution-ary algorithms. IEEE Trans. Evol. Comput. **4**(1), 1–15 (2000)
21. Tavares, J., Pereira, F.B., Costa, E.: Multidimensional knapsack problem: a fitness landscape analysis. IEEE Trans. Syst. Man Cybern. Part B (Cybern.) **38**(3), 604–616 (2008)
22. Vose, G.E.L.M.D.: Deceptiveness and genetic algorithm dynamics. In: Foundations of Genetic Algorithms 1991 (FOGA 1), vol. 1, p. 36 (2014)
23. Yossi, B., Poli, R.: Information landscapes and the analysis of search algorithms. In: Proceed-ings of the 7th Annual Conference on Genetic and Evolutionary Computation, pp. 1287–1294. ACM (2005)

Part III
Evolutionary Algorithms for Complex Networks

Chapter 5
Evolutionary Community Detection Algorithms

In the real world, many systems can be modeled as networks where nodes represent objects and edges represent relationships between objects. The analysis of networks gives us a lot of help in understanding the features of these systems. In a network which is not regular like, the degree distribution of edges may be highly concentrated within special groups of nodes and lowly concentrated between these groups. This feature is named as community structure [9].

Community structure can be easily found in various kinds of networks like social networks, biological networks, and information networks. There are also many concrete applications, such as recommendation systems in business and task allocation in computer science. Therefore, the study of community structure has become a popular topic in recent years. The problem of detecting community structures in networks is named as community detection problems (CDPs). Many methods for CDPs have been proposed. In this chapter, we concentrate on detection methods based on EAs.

5.1 Community Detection Problems

Given a graph $G = (V, E)$, a community, also called clusters or modules, of the graph is a subgraph C with n_C nodes. All the edges connecting with the nodes in C can be divided into two categories. For a node v in C, the edges connecting v to other nodes in C are internal edges and the edges connecting v to the rest of the graph are external edges. The internal and external degree of node v, labeled as k_v^{int} and k_v^{ext}, are the numbers of internal and external edges connecting to v. Based on this, the internal and external degree of community C is defined as follows:

$$k_C^{\text{int}} = \sum_{v \in C} k_v^{\text{int}}, \quad k_C^{\text{ext}} = \sum_{v \in C} k_v^{\text{ext}} \tag{5.1}$$

© Springer Nature Switzerland AG 2019
J. Liu et al., *Evolutionary Computation and Complex Networks*,
https://doi.org/10.1007/978-3-319-60000-0_5

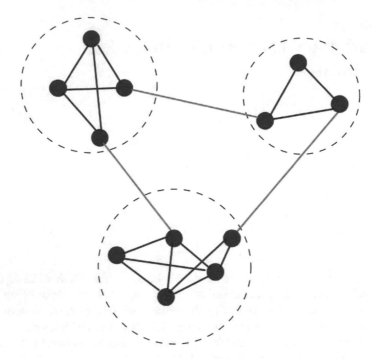

Fig. 5.1 A graph with three separated communities, enclosed by the dashed circles

The objective of CDPs is to find good communities in a graph. However, a precise definition does not yet exist for what kind of communities are good. Generally, nodes have dense connections within the community and sparse connections between communities (Fig. 5.1).

5.1.1 Community Structure

Based on whether a node in a graph can be grouped into one community or more, the community structure can be divided into two categories, named as separated community structures and overlapping community structures, respectively. The separated community structure is a partition of a given graph, where each node belongs to one and only one community. Figure 5.2 shows an example of a separated community structure [8].

The total number of possible partitions of a graph with n nodes is the nth Bell number B_n [2], which is defined by a recurrence relation equation in (5.2).

$$B_0 = 1, \quad B_{n+1} = \sum_{k=0}^{n} \binom{n}{k} B_k \tag{5.2}$$

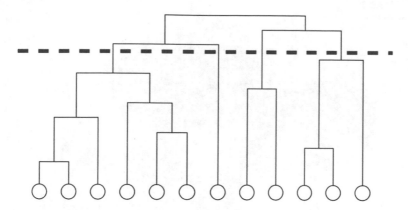

Fig. 5.2 A hierarchical tree of a graph with twelve nodes

In the limit of large n, B_n has the asymptotic form [27].

$$B_n \rightarrow \frac{1}{\sqrt{n}}[\lambda(n)]^{n+1/2}e^{\lambda(n)-n-1} \tag{5.3}$$

where $\lambda(n) = e^{W(n)}$, $W(n)$ being the Lambert W function [39]. From (5.3), we can tell that the total number of possible partitions grows faster than exponentially with the graph size n, so brutal search of all partitions of a graph is impossible when the graph is large.

The separated community structure can be hierarchically ordered if the graph has different levels at different scales. For example, students in a social network can be divided into groups according to their schools, while in each school, they can be also divided into a lower-level group according to their classes. The hierarchical structure can be represented in a tree structure like Fig. 5.2 [33]. The leaves at bottom of the tree correspond to the nodes in the graph and join together to form a larger community as we move upward. A cut at any level of the tree (the red dotted line in this case) gives the community structure at that scale.

In many real scenarios, the nodes in a network may belong to more than one group. In that case, the separated community structure will not reveal the common nodes in communities. Figure 5.3 shows an example of an overlapping community structure [35]. The overlapping nodes shown by the red big dots in the graph belong to more than one community.

The fact that a node may belong to many communities enormously increases the number of possible solutions with respect to the separated community structure. Therefore, searching for overlapping community structure is usually more computationally demanding than separated community structure.

Fig. 5.3 Overlapping community structure. Four communities are in different colors, and the overlapping nodes are represented by big red dots

5.1.2   6.1.2 Communities on Different Types of Networks

Many works on community detection problems were originally proposed for undirected unweighted networks. However, the real-world systems are more complicated. Here we introduce the adaptation of communities on different types of networks.

The relationships between elements in the real world can have a precise direction, which results in a directed network. An edge from u to v in a directed network does not imply an edge from v to u. In most community detection methods, the density within and between communities are independent of directions. Therefore, it is customary to neglect the direction of edges and consider the network as undirected. However, the directions may give extra information to improve the quality of community structure. Moreover, in some instances, neglecting the directions may lead to strange results [20, 41], which cannot reveal the real structure of systems. The adaptation of community detection methods to directed networks is a hard task. Only a few techniques can be easily adapted and make a full use of the direction information. Otherwise, the problem needs to be formulated from scratch.

The edges in networks may be associated with a value named as weight. In most cases of community detection, the weight of an edge is the measure of strength of the relationship between the two joint nodes. Therefore, many community detection methods can be easily adapted to weighted networks by updating the concept degree

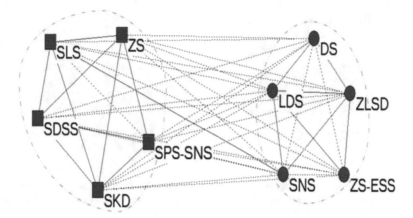

Fig. 5.4 Community structure of the Slovene Parliamentary Party Network. Two communities are shown in different colors and shapes

to sum of weights. At this point, the unweighted graph can be considered as a special case of weighted graph in which all the edges have a weight of 1.

Another important special case of weighted networks is signed networks. In signed networks, edges are in two opposite types, such as friends and foes in social networks. Signed networks can be represented by a graph with positive weighted and negative weighted edges. Figure 5.4 is an instance of signed networks, namely the Slovene Parliamentary Party Network [16, 23]. The solid edges are positive, while the dashed edges are negative.

Kunegis et al. in [17] pointed out that the negative edges can add a lot of useful information for networks. The difference between positive and negative edges asks new community detection technologies. The community detection on signed network requires more positive edges within communities and more negative edges between communities. In signed networks, the sum of weight of some elements may be zero or negative, which may results in some meaningless solutions if we take the signed network as a normal weighted network. Moreover, the two above-mentioned requirements may not be accomplished at the same time, which is shown in Fig. 5.5 and [23].

5.1.3 Measures for Evaluating Community Structure

5.1.3.1 Normalized Mutual Information

To evaluate the quality of community structure, as well as the performance of a community detection method, lots of measures have been proposed. If a real community structure is given, we can evaluate the community structure provided by our methods

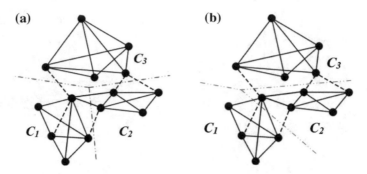

Fig. 5.5 Two possible community structures of a signed network (solid lines are positive and dashed lines are negative): **a** The partition which has more positive edges within communities, and **b** the partition which has more negative edges between communities

according to the similarity between the given structure and the output structure. For the separated community structure, the normalized mutual information (NMI) [4] can be used.

A separated community structure can be represented as a vector X of n community labels. If the ith element in X is k, then the ith node in the graph belongs to community C_k. Considering X as a random variable, the probability distribution of X and the joint distribution of X and Y are defined as follows:

$$
\begin{aligned}
P(X = k) &= |C_k^X|/|V| \\
P(X = K, Y = l) &= |C_k^X \cap C_l^Y|/|V|
\end{aligned}
\tag{5.4}
$$

where $|C_k^X|$ is the number of nodes in C_k in X and $|V|$ is the number of all nodes in the graph. The entropy of X is defined as follows:

$$
H(X) = \sum_{k \in X} h(P(X = k))
\tag{5.5}
$$

where $h(p) = -p \log(p)$, the normalized mutual information of X and Y is defined as follows:

$$
NMI(X; Y) = -\frac{H(X, Y) - H(X) - H(Y)}{H(X) + H(Y)/2}
\tag{5.6}
$$

A large *NMI* denotes that two community structures are similar. If $NMI(X; Y) = 1$, then the structures X and Y are exactly the same.

To compare two overlapping community structures, Lancichinetti et al. [18] proposed a generalized NMI (GNMI). In overlapping community structure, the kth community can be represented by a binary vector of n elements X_k. The ith node belongs to the kth community if the ith element in X_k is 1. All these communities consist of the overlapping community structure **X**.

The probability distribution of random variables from X_k is given as follows:

$$P(X_k = 1) = |C_k^x|/|V|, \quad P(X_k = 0) = 1 - |C_k^X|/|V| \tag{5.7}$$

and the joint distribution of X_k and Y_l is

$$
\begin{aligned}
P(X_k = 1, Y_l = 1) &= |C_k^X \cap C_l^Y|/|V| \\
P(X_k = 1, Y_l = 0) &= \left(|C_k^X| - |C_k^X \cap C_l^Y|\right)/|V| \\
P(X_k = 0, Y_l = 1) &= \left(|C_l^Y| - |C_k^X \cap C_l^Y|\right)/|V| \\
P(X_k = 1, Y_l = 1) &= |C_k^X \cap C_l^Y|/|V|
\end{aligned}
\tag{5.8}
$$

In particular, we can define the normalized conditional entropy of X_k with respect to all the communities of $\mathbf{Y}$ as

$$H(X_k|\mathbf{Y})_{\text{norm}} = \frac{1}{H(X_k)} \min_{Y_l \in \mathbf{Y}} H(X_k|Y_l) \tag{5.9}$$

The definition tries to choose Y_l similar to X_k by minimizing the conditional entropy of the two community structures. However, if Y_l is similar to the complementary of X_k, the conditional entropy is also small. To overcome this issue, (5.10) is added as the constraint of (5.9). If no Y_l satisfies the constraint, $H(X_k|\mathbf{Y}) = 1$.

$$
\begin{aligned}
h(P(X_k = 1, Y_l = 1)) &+ h(P(X_k = 0, Y_l = 0)) \\
&> h(P(X_k = 1, Y_l = 0)) + h(P(X_k = 0, Y_l = 1))
\end{aligned}
\tag{5.10}
$$

For all the communities in $\mathbf{X}$, the normalized condition entropy is

$$H(\mathbf{X}|\mathbf{Y})_{\text{norm}} = \frac{1}{|\mathbf{X}|} \sum_{X_k \in \mathbf{X}} H(X_k|\mathbf{Y}) \tag{5.11}$$

where $|X|$ is the number of communities in structure $\mathbf{X}$. The definition of GNMI is

$$\text{GNMI}(\mathbf{X}; \mathbf{Y}) = 1 - \frac{1}{2}(H(\mathbf{X}|\mathbf{Y})_{\text{norm}} + H(\mathbf{Y}|\mathbf{X})_{\text{norm}}) \tag{5.12}$$

NMI is proposed for the comparison between two community structures. To evaluate the quality of a community structure, a known best community structure should be given, which is a seldom the case in practical applications. Therefore, NMI and GNMI are mainly used in evaluating the performance of community detection algorithms on some benchmark data.

5.1.3.2 Modularity

Newman et al. [32] proposed the modularity, usually labeled as Q, to evaluate the quality of community structure. The measure has a good performance and soon became the most popular measure in the field of community detection. Moreover, Q is based on the idea that a random graph is not expected to have community structure. Therefore, the difference between density with communities on the original network and a random network can reveal the quality of community structure. The modularity is defined as follows:

$$Q = \frac{1}{2M} \sum_{ij} (A_{ij} - P_{ij}) \delta(C_i, C_j) \tag{5.13}$$

where m is the total number of edges in the graph, A is the adjacent matrix, P_{ij} is the expected number of edges between nodes i and j in a random network, C_i (C_j) is the community which node i (j) belongs to, and the δ *function* yields 1 if i and j are in the same community, 0 otherwise.

The value of P_{ij} depends on the chosen of null model. It is preferable to use a null model with the same degree distribution of the original graph. With this assumption, the probability to form an edge between i and j is $k_i k_j / 4m^2$, where k_i and k_j are the degree of nodes i and j, respectively. Therefore, the final expression of modularity can be formulated as

$$Q = \frac{1}{2m} \sum_{ij} \left(A_{ij} - \frac{k_i k_j}{2m} \right) \delta(C_i, C_j) \tag{5.14}$$

The large modularity indicates a good quality of community structure. The modularity value is always less than one and may be negative. The modularity of a community structure that takes the whole network into one community is zero.

The modularity can be easily adopted on directed and weighted networks. The modularity of a community structure on a directed network [20] is

$$Q_D = \frac{1}{m} \sum_{i,j} \left(A_{ij} - \frac{k_i^{out} k_j^{in}}{m} \right) \delta(C_i, C_j) \tag{5.15}$$

where k_i^{out} is the number of out edges from i and k_j^{in} is the number of in edges to j.

For weighted but unsigned networks, the degrees in (5.14) change to the sum of weights, resulting in (5.16),

$$Q_W = \frac{1}{2w} \sum_{ij} \left(w_{ij} - \frac{w_i w_j}{2w} \right) \delta(C_i, C_j) \tag{5.16}$$

where w_ij denotes the weight of the edge between i and j, w_i (w_j) denotes the sum weight of all edges connected with i (j), and w denotes the sum weight of all edges in the network.

For the signed case, the sums of weights of positive edges and negative edges need to be calculated separately. Gómez et al. [12] proposed the modularity for signed network community detection as follows:

$$Q_s = \frac{1}{2w^+ + 2w^-} \sum_{i,j} \left(w_{ij} - \left(\frac{w_i^+ w_j^-}{2w^+} - \frac{w_i^- w_j^-}{2w^-} \right) \right) \delta(C_i, C_j) \qquad (5.17)$$

where w_i^+ (w_j^+) denotes the absolute value of sum weight of all positive edges connected with i (j), w_i^- (w_j^-) denotes the absolute value of sum weight of all negative edges connected with i (j), w^+ and w^- are the absolute values of sum weight of all positive and negative edges in the network, respectively.

Shen et al. [43] extended the modularity to overlapping community structure. The definition of modularity of overlapping community structure is

$$Q_o = \frac{1}{2m} \sum_{C} \sum_{i,j \in C} \frac{1}{O_i O_j} \left(A_{ij} - \frac{k_i k_j}{2m} \right) \qquad (5.18)$$

where O_i (O_j) denotes the number of communities which node i (j) belongs to. To combine the signed network and overlapping community structure, the definition in [23] is given as follows:

$$Q_s = \frac{1}{2w^+ + 2w^-} \sum_{C} \sum_{i,j \in C} \frac{1}{O_i O_j} \left(w_{ij} - \left(\frac{w_i^+ w_j^+}{2w^+} - \frac{w_i^- w_j^-}{2w^-} \right) \right) \qquad (5.19)$$

The modularity has some limitations. A very fundamental issue proposed by Fortunato et al. in [7] is the resolution limitation. In their work, they pointed out that when the natural communities are smaller than a certain scale depending on the size of networks, the optimum Q does not reveal the natural structure. Figure 5.6 shows an example of the resolution limitation of modularity [7].

When the network is large and the degrees of nodes are small, the expected number of edges between can be very small. As a result, a few existing edges may increase Q largely and two adjacent community may be merged mistakenly. The reason for this issue is that the null model assumes that a node can build relationships with any other nodes, even it is far away. This global view of nodes is questionable, especially in real large systems.

Fig. 5.6 Resolution limit of
modularity. K_l are a clique of
l nodes. The maximum
modularity corresponds to a
partition whose clusters
include two or more cliques,
which are indicated by
dashed contours here

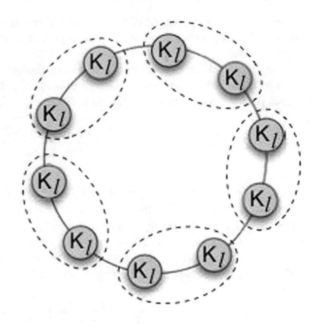

5.1.3.3 Tightness

Huang et al. in [14] proposed a new measure of quality of community structure based
on similarity of nodes. The similarity $s(u, v)$ for nodes u and v is defined as

$$s(i, j) = \frac{\sum\limits_{k \in \Gamma(i) \cap \Gamma(j)} w_{ik} \cdot w_{kj}}{\sqrt{\sum\limits_{k \in \Gamma(i)} w_{ik}^2} \cdot \sqrt{\sum\limits_{k \in \Gamma(j)} w_{kj}^2}} \tag{5.20}$$

where $\Gamma(i)$ is the set of node i and all its neighbors. Equation 5.20 suggests that if
two nodes have more common neighbors, the similarity of them is larger.

Based on the similarity, the measure of a single community C, named as tightness,
is defined as

$$T(C) = \frac{S_C^{in}}{S_C^{in} + S_C^{out}} \tag{5.21}$$

where S_C^{in} is the sum of similarities between any adjacent nodes both inside the
community C, and S_C^{out} is the sum of all the similarities between two nodes inside
and outside the community C, respectively.

Liu et al. [23] extended this concept to signed social networks. Based on the social
balance theory [47], [6], nodes in a social network tend to form into a balanced
structure. Three common patterns of signed networks are shown in Fig. 5.7 as well
as the potential relationships.

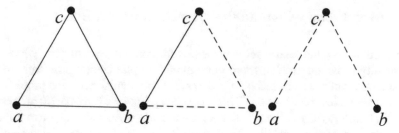

Fig. 5.7 Three patterns of relationships. Solid lines denote friend and dashed lines denotes foes. Red lines are potential relationships

Therefore, the similarity of signed networks is defined as

$$s_s(i,j) = \frac{\displaystyle\sum_{k\in\Gamma(i)\cap\Gamma(j)} \psi(w_{ik}, w_{kj})}{\sqrt{\displaystyle\sum_{k\in\Gamma(i)} w_{ik}^2} \cdot \sqrt{\displaystyle\sum_{k\in\Gamma(j)} w_{kj}^2}} \tag{5.22}$$

where

$$\psi(w_{ik}, w_{kj}) = \begin{cases} w_{ik} \cdot w_{kj} & \text{if } w_{ik} < 0 \text{ and } w_{kj} < 0 \\ 0 & \text{otherwise} \end{cases} \tag{5.23}$$

The definitions of in and out tightness of signed networks of positive and negative similarities are

$$T_{in+}(C) = \frac{S_C^{in+}}{S_C^{in+} + S_C^{out+}}, \quad T_{out-}(C) = \frac{S_C^{out-}}{S_C^{in-} + S_C^{out-}} \tag{5.24}$$

where S_C with superscript "in+" denotes the sum of positive similarities between any nodes inside C. Note that, S_C with superscript "out+" denotes the sum of all the positive similarities between nodes inside and outside C, respectively. S_C with superscript "in-" and "out-" are in a similar way except only negative similarities add up.

The tightness of signed networks is

$$T_S(C) = \frac{S_C^{in+} - S_C^{in-}}{S_C^{in+} - S_C^{in-} + S_C^{out+}} \tag{5.25}$$

5.2 Evolutionary Community Detection Algorithms

In last section, we introduced some measures for community structure, which turn community detection problems into combinatorial optimization problems. In this section, we introduce some technologies to solve community detection problems by EAs. Many studies about EAs on CDPs have been proposed. Bui et al. [3] proposed a genetic algorithm for graph partition with a schema preprocessing phase to improve GAs' space searching capability. Talbi et al. [46] proposed a parallel genetic algorithm for the graph partition problem which showed a superlinear speedup. Tasgin et al. [48] used a genetic algorithm to detect communities based on modularity. Pizzuti [37] proposed a genetic algorithm for community detection named as GA-Net using the locus-based adjacency representation and uniform crossover. It was efficient in reducing the invalid search when only the actual correlations of all nodes were considered in each operator. A new collaborative EA was proposed by Gog et al. [10] which was based on information sharing mechanism between individuals in a population. Gong et al. [13] proposed a memetic algorithm to optimize the modularity density for community detection. A local search procedure—a strategy called hill-climbing—was added to genetic algorithms which performed better than traditional GAs.

5.2.1 Representation and Operators

Encoding a community structure into an individual in EAs is a hard task since the total number of communities is unknown and the size of each community varies. Here, we introduce three widely used representations for community structure.

5.2.1.1 Directed Representation

Directed representation, also called label representation, naturally represents the community structure as a vector X of n elements (n is the number of nodes in the network). The ith node belongs to the community with label X_i. The operation on directed representation is easy to implement. Operations like crossover can swap some elements of two individuals, and mutation is changes one or more values in the vector. The evaluation of an individual is also simple since no decode step is needed.

The drawback of directed representation is its inefficiency. Since the number of communities is unknown, the maximum potential community label is n. The total number of feasible solutions is n^n, which is much larger than the Bell number. For instance, two individuals $(1, 1, 3, 2, 2, 3, 1)$ and $(4, 4, 1, 5, 5, 1, 2)$ are the representations for the same community structure. The redundancy of solution space results in poor quality and slow convergence speed. Also, the directed representation can only handle separated community detection problems. To solve overlapping community detection problem, extra technologies are necessary.

5.2.1.2 Permutation Representation

Liu et al. [26] proposed a permutation representation for community structure. This representation uses a permutation X of n nodes; i.e. two different elements in X must have different values. A decoder based on greedy search is used to convert a permutation into a community structure for further evaluation. It scans the nodes one by one according to the order of permutation and puts the nodes into one or more proper communities.

Operators for permutation representation are also a little more complex in order to maintain that the offspring is still a permutation. The partially mapped crossover method [11] and reverse mutation can be used on permutation representation.

The number of feasible solutions of permutation representation is $n!$, which is lower than directed representation but still larger than the Bell number. However, the decoding step is actually a small optimization and the evolutionary step is selecting from local optimums. Therefore, the quality of solutions is largely improved and the convergence speed is fast so that the computational complexity brought by the decoding is insignificant. Permutation representation was designed to solve both separated and overlapping community problems by adjusting the decoding scheme. The only missing part is a theoretical proof on the existence of a permutation able to be decoded into the global optimum.

5.2.1.3 Locus-Based Representation

Park et al. [36] proposed a locus-based representation. For a locus-based representation X, the ith node is in the community as the same as node X_i. It seems that the number of possible solutions is n^n, but the trick of locus-based representation is that the possible values of X_i is limited to the neighbors of ith node, because of the fact that a good community should not have an isolate node. In another word, locus-based representation is choosing some edges to form a skeleton of networks. The skeleton naturally separated into several connected components and forms communities. Figure 5.8 is an instance of locus-based representation.

The operators for locus-based representation are also very simple. Single-point and double- point crossover can be used since the offspring is also a safe locus-based representation. The mutation operator can be change X_i to another neighbor of the ith node.

A fact of locus-based representation is that a connected component decoded from the representation has one and only one loop. Therefore, merge or split communities are easy. Given a genotype X, if nodes i and k are in two different communities, labeled as C_i and C_k, respectively, we are sure about that $X_i \neq k$. Changing X_i to k has two results. One is that when i is in the loop of C_i, the loop is opened to a tree and attached to C_k to form a larger merged community. The other is that when i is not in the loop, C_i split into two parts and the part with i is merged into C_k. If nodes i and k are in the same community, changing X_i to k split the community into two smaller communities when i is not in the loop.

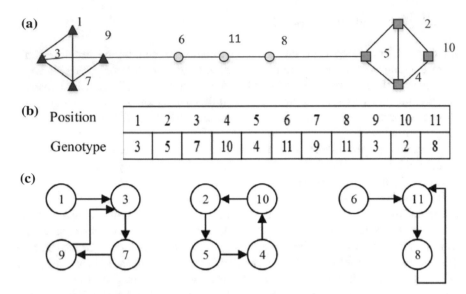

Fig. 5.8 Locus-based representation: **a** A graph with three communities, **b** locus-based representation vector, and **c** the skeleton of the graph based on **b**

The locus-based representation is simple and efficient. The decoding step needs a breadth-first search with time complexity $O(n)$. The number of possible solutions is Πk_i, which is not larger than $(m/n)^n$ (m is the number edges in networks) since $\sum k_i = m$. From Fig. 5.9, we can see that the number of possible solutions of locus-based representations increase much slower than the two other representations. However, solving overlapping community problem needs extra technologies.

5.2.2 A Multi-agent Genetic Algorithm for Community Detection

Li et al. [22] proposed a multi-agent genetic algorithm named as MAGA-Net to detect communities. Here, we first introduce MAGA-Net and then show the comparison between MAGA-Net and two other algorithms on real-life and synthetic networks.

5.2.2.1 MAGA-Net

MAGA-Net uses an agent lattice to do the searching. Therefore, we make a brief introduction of agents first. Agents have the ability to perceive and react against the environment. They have a very wide range of meanings depending on the problem that needs to be solved. In general, an agent has four properties [24, 25]: (1) it lives

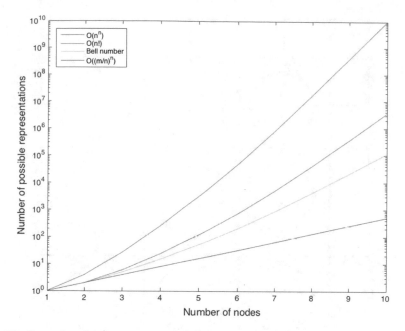

Fig. 5.9 Comparison in terms of the solution redundancy of three representations

and acts in an environment; (2) it is able to sense its local surrounding; (3) it is driven by a specific purpose; and (4) it owns some reactive behaviors. All agents compete or work together to achieve their common goals. In MAGA-Net, an agent is defined as a division of a network.

Definition 5.1 An agent is a candidate solution of the community detection problem, while the value of its energy equals the modularity defined in Eq. (5.14).

Definition 5.2 All agents live in a lattice-like environment L called agent lattice. Each agent is fixed on one point of this kind of lattice and can only exchange information with its neighbors. The number of agents is $L_{size} \times L_{size}$, and the agent lattice can be defined as the form in Fig. 5.10.

Suppose an agent locates at (m, n), $m, n = 1, 2, \ldots, L_{size}$, then its neighbors can be defined as $neighbors_{m,n}$ in Eq. 5.26. We can see $\mathbf{L}_{m,n}$'s neighbors clearly in Fig. 5.11.

$$\text{neighbors}_{m,n} = \{L_{m',n}, L_{m,n'}, L_{m,n''}, L_{m'',n}\}$$

$$
m' = \begin{cases} m - 1 & m \neq 1 \\ L_{size} & m = 1 \end{cases} \quad
n' = \begin{cases} n - 1 & n \neq 1 \\ L_{size} & n = 1 \end{cases}
$$

$$
m'' = \begin{cases} m + 1 & m \neq L_{size} \\ 1 & m = L_{size} \end{cases} \quad
n'' = \begin{cases} n + 1 & n \neq L_{size} \\ 1 & n = L_{size} \end{cases}
\tag{5.26}
$$

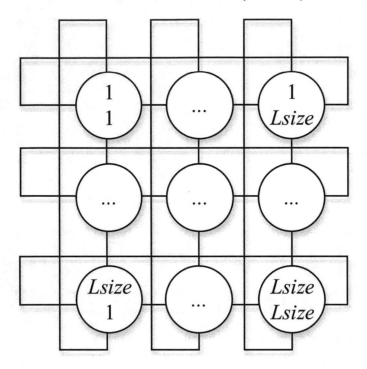

Fig. 5.10 Model of the agent lattice

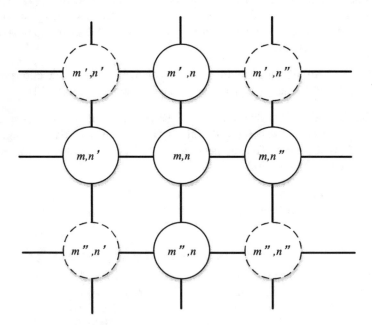

Fig. 5.11 The neighborhood of an agent

In MAGA-Net, the authors employed the locus-based adjacency representation proposed in [36] to represent agents. Besides, they also designed four genetic operators: (1) split- and merge-based neighborhood competition operators, (2) hybrid neighborhood crossovers, (3) adaptive mutation, and (4) self-learning operators. In order to gain more energy, on the one hand, all agents compete or cooperate with their neighbors through the former two operators; on the other hand, each agent has the ability to use knowledge of its neighbor or itself to get a further study using the latter two operators. All the operators are performed on $L_{m,n} = [g_1, g_2, g_3, \ldots g_N]$. $\text{Max}_m, n = [h_1, h_2, h_3, \ldots h_N]$ is $L_{m,n}$'s neighbor with maximum energy. The detail descriptions of the four operators are shown as follows.

Split- and merge-based neighborhood competition operators. The authors find that each node has one and only one link toward its adjacency, and there is one and only one loop in each community from Fig. 5.8. Starting from this point, they use $u(0, 1)$ to represent a uniformly distributed random value in the interval $[0, 1]$. If $u(0, 1) \geq 0.5$, they randomly select a gene g_i of an agent as the target gene and transform its value to one of its allele k in adjacency list when their cluster labels are different. They set s_1 as the community label of g_i, and s_2 as the allele k's label, then $s_1 \neq s_2$. There are two possible outcomes, one is that communities s_1 and s_2 merge together to form a larger new cluster; the other is that community s_1 splits into two small communities and one of them without a loop combines with the community s_2; thus, two communities are generated— one is smaller, and the other is larger. This occurs when the target gene is out of the loop of the community to which it belongs; if $u(0, 1) < 0.5$, they just transform the value of the selected gene to an arbitrary allele of its adjacency list.

Using the above idea, they design a split- and merging-based neighborhood competition operator. For each agent $L_{m,n}$ on the lattice, do a comparison between agents $L_{m,n}$ and $Max_{m,n}$. If $Energy(L_{m,n}) > Energy(Max_{m,n})$, $L_{m,n}$ is the optimal one and can survive; otherwise, agent $L_{m,n}$ will be replaced by $\text{Max}_{m,n}$ with the strategy they proposed.

Strategy: First, assign $\text{Max}_{m,n}$ to New $= [r_1, r_2, r_3, \ldots, r_N]$. Then, if $u(0, 1) \geq 0.5$, randomly select an gene r_i in New and replace it with one of its alleles k in the adjacency list under the precondition of nodes i and k are in different communities; if $u(0, 1) < 0.5$, just select an gene r_i and change its value to one of its neighbors in the adjacency list. Finally, $L_{m,n}$ is replaced by New.

Hybrid neighborhood crossover: Traditionally, a two-point crossover operator is conducted on two chromosomes, randomly selects two crossing points, and then exchanges the genes of the two chromosomes between the two points. Differing from the two-point crossover operator, the uniform crossover exchanges each gene of the two chromosomes with the same probability. These two crossover operators are both easy to implement and could be applied to MAGA-Net because of the flexible of the locus-based adjacency representation. But the main disadvantage of these two operators is that their recombination is random and you cannot guarantee to generate offspring better than the parents.

In MAGA-net, they used a hybrid neighborhood crossover operator. An agent $L_{m,n}$ on the lattice will only cross with $\text{Max}_{m,n}$ so as to obtain useful information from its neighbors and avoid random recombination. Moreover, they mix a two-point crossover operator and a uniform crossover together to provide more possibility of changes. In order to protect the good pattern, they will not change $Max_{m,n}$

For each agent $L_{m,n}$, if $u(0, 1) < p_c$, and Energy$(L_{m,n}) <$ Energy$(Max_{m,n})$, they will replace $L_{m,n}$ by mixing $L_{m,n}$ and $Max_{m,n}$ with two strategies decided by p_s. That is, if $u(0, 1) p_s$, strategy 1 is performed; otherwise, strategy 2 is selected.

Strategy 1: Randomly select two points k_1, k_2, if $u(0, 1) < 0.5$, then assign genes between k_1 and k_2 of $Max_{m,n}$ to $L_{m,n}$; otherwise, the rest genes are assigned to $L_{m,n}$.

Strategy 2: The general uniform crossover operator is conducted on $Max_{m,n}$ and $L_{m,n}$, then the newly generated agent will replace $L_{m,n}$.

Adaptive mutation: In order to avoid the useless discovery of the search space and keep the diversity of the population, they apply neighbor-based mutation operator proposed in [38]. In this operator, for each gene of an agent, if $u(0, 1) < p_m$, then the gene's value is changed to the allele of one of its neighbors.

An adaptive mutation operator changing p_m to find a better result proposed in [31] is also adopted. The mutation probability will increase with the growing number of generations when no improvement has reached. Higher p_m represents more changes of genes in each agent which makes it more likely to jump out of local optima. N_s is the termination criterion and MAGA-Net will stop when it has run N_s generations without improvement. They set t to stand for the number of generations without improvement, then p_m is showed as follows:

$$p_m' = (t/N_s + 1)p_m \tag{5.27}$$

Self-learning operator: An agent can use its knowledge to get more energy, so a local search method is essentially required to conduct on excellent agents. They generate a small-scale agent lattice sL with the size $sL_{size} \times sL_{size}$ based on Eq. 5.28. $L_{m,n}$ is the agent that will be conducted the self-learning operator.

$$(6.6)sL = \begin{cases} L_{m,n} & m' = 1 \text{ and } n' = 1 \\ sL_{m',n'} & \text{otherwise} \end{cases} \tag{5.28}$$

where $sL_{m',n'}$ is generated by conducting the neighbor-based mutation operator on $L_{m,n}$ with probability sP_m. The split- and merge-based neighborhood competition operator is conducted on sL in each generation. sN_s is the termination criterion which indicates that the self-learning operator will stop when it runs sN_s generations without improvement. At last, the best agent with maximum energy in this process replaces $L_{m,n}$. They conduct the self-learning operator on the best sl number of agents in L to further increase their energy. Algorithm 5.1 summarizes self-learning operators more clearly.

With integration of all these strategies, the pseudo-code of MAGA-Net is provided in Algorithm 5.2. The split- and merge-based neighborhood competition operator is used to eliminate agents with lower energy (smaller modularity value) from the lattice firstly. Then, hybrid neighborhood crossover and adaptive mutation operators are conducted on L to effectively explore the search space with probabilities P_c

and P_m and maintain the diversity of the population. Next, self-learning operator is performed on the best sl number of agents in L, which plays an irreplaceable role in improving the performance of MAGA-Net. The learning property of an agent determines whether or not the self-learning operator can conduct on it. If the learning label of an agent is *True*, it means the self-learning operator can conduct on this agent, otherwise not. When there is a change of an agent caused by the genetic operators, the learning label of this agent will be assigned as *True*, by which it can regain the opportunity of self-learning.

Algorithm 5.1 Self-learning Operator

Input:
  sL_t: the agent lattice of the tth generation of sL;
  sN_s: the maximum number of generations without improvement.
  $L_{m,n}$: an agent in L implemented self-learning operator;
  $sBest^t$: the most optimal agent in $sL_0, sL_1, \ldots, sL_t$;
  $sCBest^t$: the best agent in sL_t;
  sP_m: Mutation probability;
Output:
  $L_{m,n} \leftarrow sBest^t$;
  $Learning(L_{m,n}) \leftarrow False$;

$t \leftarrow 0$;
$n \leftarrow 0$;
$\mathbf{sL_0} \leftarrow$ initialize sL using the neighbor-based-mutation-operator according to Eq. 5.28 with probability sP_m and update $sBest^0$;
while $n < sN_s$ **do**
  $t \leftarrow t + 1$;
  $\mathbf{sL_t} \leftarrow$ conduct the split-and-merging-based-neighborhood-competition-operator on sI_t;
  Update $sCBest^t$;
  if $Energy(sCBest^t) > Energy(sBest^{t-1})$ **then**
    $n \leftarrow 0$;
    $sBest^t \leftarrow sCBest^t$;
  else
    $n \leftarrow n + 1$;
    $sBest^t \leftarrow sBest^{t-1}$;
    $sCBest^t \leftarrow sBest^t$;
  end if
end while

Algorithm 5.2 MAGA-Net

Input: L_t: the agent lattice of the tth generation of L;
  sL: the number of agents carried out self-learning;
  $Best^t$: the best agent in $L_0, L_1, \ldots, L_t$;
  $CBest^t[l]$: the best sl agents of L_t;
  $CBest^t$: the best agent in $CBest^t[l]$;
  P_c: Crossover probability;
  P_m: Mutation probability;
  P_s the probability of deciding which strategy will be chosen in hybrid neighborhood crossover;
  N_s: the maximum number of generations without improvement;
Output: Transform the best agent in L_t into a partition solution and output.

$t \leftarrow 0$;
$n \leftarrow 0$;
$\mathbf{L_0} \leftarrow$ initialize the population by locus-based adjacency representation, assign the Learning labels of L_0 as *True* and update $Best^0$;
while $n < N_s$ **do**
  $t \leftarrow t + 1$;
  $\mathbf{L_t} \leftarrow$ conduct the split-and-merging-based-neighborhood-competition-operator on L_t and update Learning labels of L_t;
  $\mathbf{L_t} \leftarrow$ conduct the hybrid-neighborhood-crossover-operator on L_t with probabilities P_c and P_s and update Learning labels of L_t;
  $\mathbf{L_t} \leftarrow$ conduct the adaptive-mutation-operator on L_t with probability P_m and update Learning labels of L_t;
  $CBest^t[sl] \leftarrow$ Finding the best sl agents of $\mathbf{L_t}$;
  for $i = 1$ **to** sl **do**
    if $Learning(CBest^t[i]) == True$ **then**
      Conduct the self-learning-operator on $CBest^t[i]$;
    end if
  end for
  Update $CBest^t$;
  if $Energy(CBest^t) > Energy(Best^{t-1})$ **then**
    $n \leftarrow 0$;
    $Best^t \leftarrow CBest^t$;
  else
    $n \leftarrow n + 1$;
    $Best^t \leftarrow Best^{t-1}$;
    $CBest^t \leftarrow Best^t$;
  end if
end while

5.2.2.2 Experiments on Real-Life and Synthetic Networks

Li et al. [22] tested the performance of MAGA-Net on four well-known real-world networks and large-scale synthetic LFR networks. They also made systematic comparisons between MAGA-Net and two representative algorithms, namely GA-Net [37] and Meme-Net [13]. Besides, *NMI* [4] defined in Eq. 5.6 was employed to calculate the similarity between the detected partitions and the true ones.

Results on four real-world networks, namely the karate club network [52], the dolphins network [28], the political books network [15], and the college football network [9] are shown in Table 5.1. BKR is the best-known results of the four networks from [1, 29, 34, 49, 51]. The average, maximum, and standard deviations (Q_{Avg}, Q_{Max}, Q_{Std}) of modularity Q and the times taken by MAGA-Net are reported in Table 5.1.

Table 5.1 shows that MAGA-Net outperforms the two other algorithms on all networks. Under the same number of evolutions and experiment environment, MAGA-Net can quickly converge to the optimal solution with a good stability. Moreover, compared to BKR, MAGA-Net has the ability to reach all the best-known results with approximating as for the values of BKR are approximations themselves. Figure 5.12 presents the clear community structure found by MAGA-Net for the four famous real-world networks.

Figure 5.13 plotted the comparison of three algorithms with respect to NMI. As can be seen, partitions obtained by MAGA-Net are more similar to the real ones than those obtained by the two other algorithms.

LFR networks [19] also used to test community discovery methods' performance. Here, six different networks were generated as u ranging from 0.1 to 0.6 with size 1,000 and 5,000, respectively. u is the mixing parameter, and it stands for the proportion of edges besides communities. The communities become harder to detect as the value of u growing. The comparisons of average value of NMI of three algorithms are presented in Figs. 5.14 and 5.15.

As shown in Fig. 5.14, MAGA-Net has a good performance in the LFR networks with 1,000 nodes. The value of NMI changes from 1 to 0.8 with u ranging from 0.1 to 0.6, which implies that the communities detected by MAGA-Net have a high

Table 5.1 The comparison in terms of modularity on the four real-world networks

Networks	BKR	GA-Net			Meme-Net			MAGA-Net			Times (s)
		Q_{Avg}	Q_{Max}	Q_{Std}	Q_{Avg}	Q_{Max}	Q_{Std}	Q_{Avg}	Q_{Max}	Q_{Std}	
Karate	0.420	0.3741	0.4198	0.0767	0.4083	0.4198	0.0134	0.4194	**0.4198**	0.0019	0.0202
Dolphins	0.529	0.4928	0.5227	0.0119	0.4273	0.5025	0.3050	0.5271	**0.5286**	0.0007	0.0733
Political books	0.527	0.4871	0.5212	0.0369	0.4436	0.5139	0.0216	0.5270	**0.5273**	0.0001	0.2680
Football	0.605	0.5020	0.5561	0.0237	0.4904	0.5492	0.0233	0.6020	**0.6046**	0.0026	0.3776

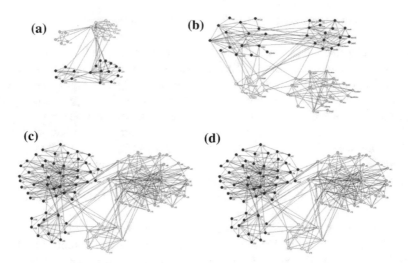

Fig. 5.12 Community structure found for the four well-known benchmark networks by MAGA-Net: **a** karate club network, $Q = 0.4198$, **b** dolphins network, $Q = 0.5286$, **c** political books network, $Q = 0.5273$, and **d** college football network, $Q = 0.6046$

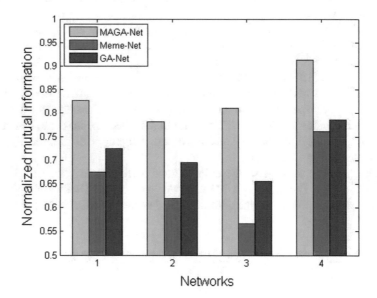

Fig. 5.13 The comparisons of average value of NMI obtained by MAGA-Net, Meme-Net and GA-Net on the four benchmark networks: (1) karate club network, (2) dolphins network, (3) political books network, and (4) college football network

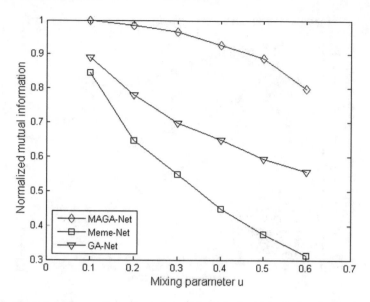

Fig. 5.14 The comparisons of average value of NMI obtained by MAGA-Net, Meme-Net and GA-Net on synthetic LFR networks as u ranging from 0.1 to 0.6. The number of nodes is now $N = 1,000$

similarity with the real ones. However, the limitations of Meme-Net and GA-Net are exposed as the size of networks and the values u increases.

The experimental results in Fig. 5.15 show that Meme-Net and GA-Net cannot handle large networks with 5,000 nodes, while MAGA-Net obtains good results with the size 5,000. Changing the value of u from 0.1 to 0.6 makes it much harder to detect communities, but MAGA-Net overcomes the difficulties and still has a good performance which comprehensively verifies the effectiveness of MAGA-Net.

All the above results show that MAGA-Net has a good performance. In terms of speed, MAGA-Net can converge to the global optimal solution with a small number of evolutions, while Meme-Net and GA-Net are far from convergence with the same number of evolutions. According to the common sense, Meme-Net with a local search should be better than GA-Net however, all the results show that GA-Net has a relatively good performance. The main reason is that the local search method in Meme-Net wastes a lot of computational resources which result in no improvement of the algorithm. This also verifies the superiority of MAGA-Net from the opposite side.

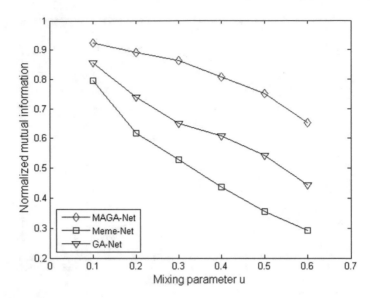

Fig. 5.15 The comparisons of average value of NMI obtained by MAGA-Net, Meme-Net, and GA-Net on synthetic LFR networks as u ranging from 0.1 to 0.6. The number of nodes is now $N = 5,000$

5.3   Multi-objective Evolutionary Algorithms for Community Detection

Although many community detection (CD) methods have been proposed, most of them can only handle networks without negative links, namely unsigned networks. However, many complex systems in social world can be modeled as networks with both positive and negative links, namely signed networks (SNs). In fact, SNs have been widely used to represent many types of social relationships. In addition to social networks, many biological networks are also signed. For instance, the interactions between genes in gene regulatory networks can be enhanced or repressed. Therefore, SNs can represent more general relationships between individuals in social or biological networks. Positive links denote friendship, cooperation, trust, enhancement, etc., while negative links denote hostility, dislike, distrust, repression, etc.

Due to the importance of community structure as a topological property of social networks, methods that can detect communities from SNs are hardly needed. In unsigned networks, community structure is defined as a group of nodes or vertices that have dense connections within groups and sparse connections between groups, whereas for SNs, communities are defined not only by the density of links but also by the signs of links. That is, within communities, the links should be positive and dense, and between communities, the links should be negative or positive and sparse. But this problem is by no means straightforward since it is natural to have some negative links within groups and, at the same time, some positive links between groups. Also, nodes connected by positive links do not belong to the same community, either. Thus, more robust community partitions should properly disregard and retain some positive and negative links so as to identify more natural communities [50]. To handle these situations, a multi-objective evolutionary algorithm is applied to community detection problem.

5.3.1 MEAs-SN

Pizzuti in [38] proposed a multi-objective genetic algorithm to uncover community structure, which optimized two objective functions, respectively, for maximizing connections within the same community and minimizing connections between different communities. Shi et al. in [44] formulated a multi-objective framework for CD and proposed an MOEA for finding efficient solutions under the framework. Shi et al. in [45] also analyzed the correlations of 11 objective functions that have been used or can potentially be used for CD, and the results showed that MOEAs optimizing over a pair of negatively correlated objectives usually performed better than the single-objective algorithm optimizing over either of the original objectives. Liu et al. in [26] proposed a MOEA to detect separated and overlapping communities simultaneously.

MEA$_s$-SN is a multi-objective evolutionary algorithm for community detection based on a MOEA/D framework [23]. The method can solve both separated and overlapping community detection problem on signed and unsigned networks. Here we will give a brief introduction of MEA$_s$-SN.

5.3.1.1 Direct and Indirect Combined Representation

The representations have a great effect on both the evolutionary operators can be used and the overall efficiency of the resulting EAs. Usually, in the field of EAs, representations fall into two broad categories: direct and indirect. A *direct* representation is the natural representation and can be evaluated easily. An *indirect* representation is not complete in itself, and a *decoder* that transforms the solution in the indirect representation into one in the direct representation is required.

In existing literature for CD based on EAs, the character string representation [48] and the locus-based adjacency representation [44] are usually used. These are two direct representations and can be evaluated easily. To detect separated and overlapping communities simultaneously, an indirect representation is proposed in [4]. Since the decoder implies a heuristic search, this indirect representation can find better candidate solutions. However, since the decoder needs to be executed before evaluating each individual, the computational cost is high. To overcome this defect, we designed a direct and indirect combined representation based on the character string representation and the indirect representation proposed in [4]. In this new combined representation, each individual is defined as follows:

Definition 5.3 An individual, **A**, consists of two components. The first component is a permutation of all nodes in V, labeled as $\mathbf{A}\langle\mathbf{P}\rangle$

$$\mathbf{A}\langle\mathbf{P}\rangle = \{v_{\pi_1}, v_{\pi_2}, \ldots, v_{\pi_n}\} \tag{5.29}$$

where $(\pi_1, \pi_2, \ldots, \pi_n)$ is a permutation of $(1, 2, \ldots, n)$. The second component is a vector with n elements, labeled as $\mathbf{A}\langle\mathbf{C}\rangle$

$$\mathbf{A}\langle\mathbf{C}\rangle = (c_1, c_2, \ldots, c_n) \tag{5.30}$$

where c_i, $1 \le i \le n$ denotes that node v_i belongs to community C_{c_i}

Clearly, $\mathbf{A}\langle\mathbf{P}\rangle$ is the indirect representation part and a decoder is required to transform it to the actual community structure. $\mathbf{A}\langle\mathbf{C}\rangle$ is the character string representation. Nodes v_i and v_j are in the same community if $c_i = c_j$.

Let $\mathbf{C} = \{C_1, C_2, \ldots, C_m\}$ be a set of communities and the decoder first initializes $\mathbf{C}$ to empty. Then, according to the order of $\mathbf{A}\langle\mathbf{P}\rangle$, for each node, check whether this node can increase $T_S(C)$; that is, whether (5.31) can be satisfied,

$$T_S(C_j \cup v_i) > T_S(C_j) \tag{5.31}$$

where C_j is one of existing communities and v_i is one node in $\mathbf{A}\langle\mathbf{P}\rangle$. For detecting separated communities, as long as one community is found to satisfy (5.31), the process is stopped and this node is added to this community. For detecting overlapping communities, this node is added to all communities that satisfy (5.31). If no existing community satisfies (5.31), this node itself forms a new community and this new community is added to $\mathbf{C}$. Further, for detecting overlapping communities, two communities with more than half identical nodes (i.e., the condition in (5.32) is satisfied) are merged.

$$\forall C_i \ne C_j, \quad \frac{|C_i \mid C_j|}{|C_i|} > \frac{1}{2} \quad or \quad \frac{|C_i \mid C_j|}{|C_j|} > \frac{1}{2} \tag{5.32}$$

The two objective functions of $\text{MEA}_s\text{-SN}$ are based on the signed tightness in Eq. 5.33. For a community structure $\mathbf{C}$ consisted of m communities labels as $C_1, C_2, \ldots, C_m$, the objective functions are

$$\begin{cases} \max f_{in+}(\mathbf{C}) = \dfrac{1}{m} \sum_{i=1}^{m} T_{in+}(C_i) \\[3mm] \max f_{out-}(\mathbf{C}) = \dfrac{1}{m} \sum_{i=1}^{m} T_{out-}(C_i) \end{cases} \tag{5.33}$$

To handle both overlapping and separated community structure, $\text{MEA}_s\text{-SN}$ uses directed representation and permutation representation. The decoder of permutation representation is given in Algorithm 5.3.

Algorithm 5.3 The decoder

Input: $A\langle P\rangle = \{v_{\pi_1}, v_{\pi_2}, \ldots, v_{\pi_n}\}$;
Output: $C = \{C_1, C_2, \ldots, C_m\}$;
1: $C \leftarrow \varnothing$;
2: **for** $i = 1$ **to** n **do**
3:    **for** $j = 1$ **to** $|C|$ **do** /*$|C|$ *denotes the number of communities in* C*/
4:      **if** C_j and v_{π_i} satisfy (17) **then**
5:        $C_j \leftarrow C_j \bigcup v_{\pi_i}$ and update $T_{signed}(C_j)$;
6:      **end if**
7:      **if** detect separated communities **then**
8:        break;
9:      **end if**
10:    **end for**
11:    **if** v_{π_i} has not been added to any community **then**
12:      $C \leftarrow C \bigcup \{v_{\pi_i}\}$;
13:    **end if**
14: **end for**
15: **if** detect overlapping communities **then**
16:    **while** there are two communities in C satisfying (5.32) **do**
17:      Merge these two communities;
18:    **end while**
19: **end if**

5.3.1.2 Evolutionary Operators

For individuals represented by the direct and indirect combined coding, evolutionary operators can be conducted on both $A\langle P\rangle$ and $A\langle C\rangle$. For $A\langle P\rangle$, the *Partially Matched Crossover* (PMX) proposed in [11] is employed. This operator was designed for the representation of array of individuals to solve traveling salesman problems. In addition, a mutation operator, which randomly selects two elements in the permutation to swap, is also used.

For $A\langle C\rangle$, the one-way crossover operator introduced in [6] is employed, and a new tightness-based mutation operator is designed as follows. Let $A\langle C\rangle = (c_1, c_2, \ldots, c_n)$, then for each c_i, $1 \le i \le n$, if $U(0, 1)$, which is a uniformly distributed random number in the range of [0, 1], is larger than T_{signed} of community C_{c_i}, then select a node v_j from the neighbors with positive similarities of v_i based on the roulette wheel selection according to their similarity, and then assign c_j to c_i. If no neighbor has positive similarity, then v_j will be the neighbor with the largest similarity.

5.3.1.3 Implementation of MEAs-SN

MEA$_s$-SN is implemented under the framework of MOEA/D, to make use of the advantages of both $\mathbf{A}\langle\mathbf{P}\rangle$ and $\mathbf{A}\langle\mathbf{C}\rangle$, in MEA$_s$-SN, the population is first initialized to $\mathbf{A}\langle\mathbf{P}\rangle$; that is, *Popsize* permutations are randomly generated. Then, for detecting separated communities, each permutation is transformed to a set of communities by the decoder. This set of communities is further transformed to $\mathbf{A}\langle\mathbf{C}\rangle$, which can be realized easily in linear time. In the following evolutionary process, each individual is represented as $\mathbf{A}\langle\mathbf{C}\rangle$, and the corresponding operators are conducted. In this way, the initial $\mathbf{A}\langle\mathbf{P}\rangle$ population can generate a better population through the decoder, and the following operations on $\mathbf{A}\langle\mathbf{C}\rangle$ can be realized time efficiently. In this way, the algorithm can benefit from the advantages of both $\mathbf{A}\langle\mathbf{P}\rangle$ and $\mathbf{A}\langle\mathbf{C}\rangle$. For detecting overlapping communities, $\mathbf{A}\langle\mathbf{P}\rangle$ is used during the whole evolutionary process since $\mathbf{A}\langle\mathbf{C}\rangle$ cannot handle overlapping communities.

The main framework of MEA$_s$-SN is MOEA/D [53]. MOEA/D decomposes a multi-objective problem into several scalar optimization subproblems and optimizes them simultaneously. In MEA$_s$-SN, the Tchebycheff approach is employed, which is defined in Eq. 5.34.

$$g^{te}(\mathbf{C}|(\lambda_{in+}, \lambda_{out-}), (f^*_{in+}, f^*_{out-}) = \max(\lambda_{in+}|f_{in+}(\mathbf{C}) - f^*_{in+}|, \lambda_{out-}|f_{out-}(\mathbf{C}) - f^*_{out-}|) \quad (5.34)$$

where g^{te} denotes the obtained scalar objective function to minimize. $\mathbf{C}$ is the community structure. (f^*_{in+}, f^*_{out-}) is the reference point corresponding to the maximum of the two objective functions. Moreover, $(\lambda_{in+}, \lambda_{out-})$ is the weight vector that satisfies that $\lambda_{in+} \geq 0$, $\lambda_{out-} \geq 0$, and $\lambda_{in+}\lambda_{out-} = 1$. The pseudo-code of MEA$_s$-SN is given in Algorithm 5.4.

5.3.2 The Experiments of MEAs-SN

In this section, the performance of MEA$_s$-SN is validated on benchmark and synthetic networks and compared with those of three existing algorithms, namely FEC [50], the Louvain method [30, 42], and CSA$_{HC}$-SN [21]. In the following experiments, *Popsize* and *Gen* of MEA$_s$-SN are set to 100. The original literature showed that MOEA/D performs well if the neighborhood size T is greater than 3, thus T is set to 20.[1] Next, the measures used to evaluate the performance of different algorithms and a synthetic network generator are first introduced. Then, three groups of experiments are conducted, which are respectively on benchmark networks, separated and overlapping CD from synthetic SNs.

---

[1]The source codes of MEA$_s$-SN can be downloaded at http://see.xidian.edu.cn/faculty/liujing/.

Algorithm 5.4 MEA$_s$-SN

Input: $G = (V, E, w)$;
    Popsize: the size of the population;
    A uniform spread of *Popsize* weight vectors: $\left(\lambda^1_{pos-in}, \lambda^1_{neg-out}\right)$,
    $\left(\lambda^2_{pos-in}, \lambda^2_{neg-out}\right), \ldots, \left(\lambda^{Popsize}_{pos-in}, \lambda^{Popsize}_{neg-out}\right)$;
    T: the number of weight vectors in the neighborhood of each weight vector;
    Gen: the number of generations;
Output: The final population;
  1: Calculate the signed similarity of any two connected nodes;
  2: Calculate the Euclidean distances between any two weight vectors and then calculate the T closest weight vectors to each weight vector.
  3: **for** each $i = 1, 2, \ldots, Popsize$ **do**
  4:    set $B(i) = \{i_1, i_2, \ldots, i_T\}$, where $\left(\lambda^{i_1}_{pos-in}, \lambda^{i_1}_{neg-out}\right)$, $\left(\lambda^{i_2}_{pos-in}, \lambda^{i_2}_{neg-out}\right), \ldots,$
     $\left(\lambda^{i_T}_{pos-in}, \lambda^{i_T}_{neg-out}\right)$ are the T closest weight vectors to $\left(\lambda^i_{pos-in}, \lambda^i_{neg-out}\right)$
  5: **end for**
  6: Initialize the population in the form $\mathbf{A}\langle\mathbf{P}\rangle$: randomly generate *Popsize* permutations of n nodes, labeled as $\mathbf{A}\langle\mathbf{P}^1_1\rangle, \mathbf{A}\langle\mathbf{P}^1_2\rangle, \ldots, \mathbf{A}\langle\mathbf{P}^1_{Popsize}\rangle$, where the superscript denotes the current generation;
  7: **if** Detecting separated communities **then**
  8:    Decode $\mathbf{A}\langle\mathbf{P}^1_1\rangle, \mathbf{A}\langle\mathbf{P}^1_2\rangle, \ldots, \mathbf{A}\langle\mathbf{P}^1_{Popsize}\rangle$ to $\mathbf{A}\langle\mathbf{C}^1_1\rangle, \mathbf{A}\langle\mathbf{C}^1_2\rangle, \ldots, \mathbf{A}\langle\mathbf{C}^1_{Popsize}\rangle$;
  9: **end if**
10: Initialize the reference idea point $(f^*_{pos-in}, f^*_{neg-out})$;
11: **for** $i = 1$ **to** *Gen* **do**
12:   **for** $j = 1$ **to** *Popsize* **do**
13:     **if** Detecting separate communities **then**
14:       Randomly select another individual, and conduct the one-way crossover operator on this individual and $\mathbf{A}\langle\mathbf{C}^i_j\rangle$;
15:       Conduct the tightness based mutation operator on $\mathbf{A}\langle\mathbf{C}^i_j\rangle$;
16:     **end if**
17:     **if** Detecting overlapping communities **then**
18:       Randomly select another individual, and conduct the PMX crossover operator on this individual and $\mathbf{A}\langle\mathbf{P}^i_j\rangle$;
19:       Conduct the swap mutation operator on $\mathbf{A}\langle\mathbf{P}^i_j\rangle$;
20:     **end if**
21:     Update the reference idea point $(f^*_{pos-in}, f^*_{neg-out})$;
22:     Update the neighborhood solutions based on $B(i) = \{i_1, i_2, \ldots, i_T\}$;
23:   **end for**
24: **end for**

5.3.2.1 The Synthetic Network Generator

Combining the LFR benchmark with the one proposed in [50], Liu et al. [23] design a new SNs generator, which is labeled as $SRN(n, k, maxk, t1, t2, minc, maxc, on, om, \mu, P-, P+)$. Here, n is the number of nodes, and k and $maxk$ are the average and maximum degree of each node. $t1$ and $t2$ are the minus exponents for the degree and community size distributions, both of which are power laws. $minc$ and $maxc$ are the minimum and maximum community size. on and om are, respectively, the number of overlapping nodes and the number of memberships of overlapping nodes. μ is the fraction of links that each node shares with nodes in other communities, which controls the cohesiveness of the communities inside the generated SNs. Thus, the higher the value of μ is, the more ambiguous the community structure is. P_- is the fraction of negative links within communities, while P_+ is the fraction of positive links between communities. Ideally, negative links should be between communities and positive links should be within communities. Thus, P_- and P_+ are two parameters used to adjust the noise level. Being the same with μ, the larger the values of P_- and P_+ are, the more ambiguous the community structure is. As can be seen, this benchmark poses severe and flexible tests to algorithms. Given a fixed μ, we can control the noise level by adjusting P_- and P_+. In general, it will be more difficult to extract communities correctly when the values of P_- and P_+ are large. In the following experiments, the capability of MEA_s-SN in handling different μ, P_-, and P_+ are systematically tested.

5.3.2.2 Experiments on Benchmark Networks

In this subsection, four benchmark SNs widely used, including two illustrative SNs and two real social SNs, are employed to validate the performance of MEA_s-SN. Moreover, although MEA_s-SN is designed for SNs, it can also be used to unsigned networks. Therefore, three popular unsigned benchmark networks, namely the Zachary karate club network, the bottlenose dolphin network, and the American football network, are also tested.

The two illustrative SNs came from [50] and each has 28 nodes. Their topological structures are shown in Fig. 5.16. The community structures obtained by MEA_s-SN are given in Fig. 5.17. As can be seen, they can be divided into three communities. The links within communities are positive, and those between communities are negative. MEA_s-SN found the correct partitions successfully for both networks, and $NMI = 1$.

The first real social network is the Slovene Parliamentary Party Network, which is the relation network of 10 parties of the Slovene Parliamentary in 1994 [16]. Positive links mean the two parties' Parliament activities are similar, while negative links mean their activities are dissimilar. Figure 5.18a shows the original topological structure, and the community structure obtained by MEA_s-SN is shown in Fig. 5.18b. As can be seen, all parties are separated into two opponent communities. This result is the same as that given by Kropivnik and Mrvar in [16].

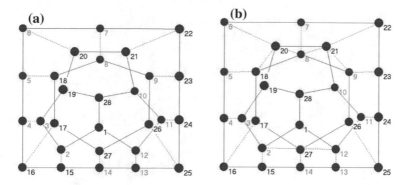

Fig. 5.16 The original topological structure of the two illustrative SNs from [22]. Solid edges denote positive links and dashed edges denote negative links

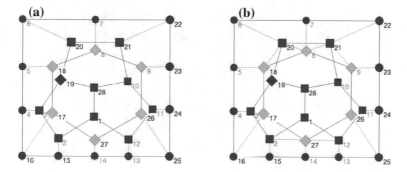

Fig. 5.17 The partitions of the two illustrative SNs obtained by MEA$_s$-SN. Nodes with the same color and shape belong to the same community

The second real social network is the Gahuku-Gama subtribes network, which was generated by Read on the cultures of highland New Guinea proposed in [40]. It regards the political alliances and oppositions among 16 Gahuku-Gama subtribes, which were distributed in a particular area and were involved in warfare with each other in 1954. Positive and negative links represent the political arrangements with positive and negative ties, respectively. Figure 5.19a shows the original topological structure, and the community structures obtained by MEA$_s$-SN are shown in Fig. 5.19b and c. As can be seen, two meaningful partitions were found. Figure 5.19b has a higher value of f_{pos-in} while Fig. 5.19c has a higher value of $f_{neg-out}$, and these results are identical to that reported in [5, 40].

To further evaluate the effectiveness of MEA$_s$-SN on unsigned networks, Fig. 5.20 presents the community structures found by MEA$_s$-SN for the Zachary karate club network, the bottlenose dolphin network, and the American football network. As can be seen, for the Zachary karate club network, the nodes are divided into two groups. For the bottlenose dolphin network, the nodes are divided into four groups. For the

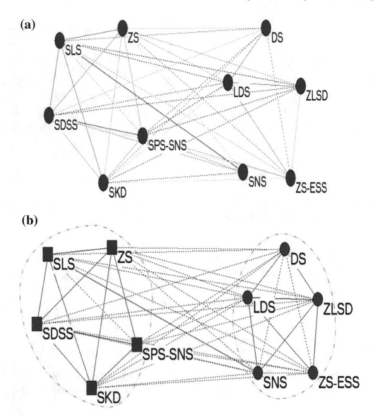

Fig. 5.18 **a** The topological structure of the Slovene Parliamentary Party Network, **b** the community structure obtained by MEA$_s$-SN

American football network, the nodes are divided into 11 groups. The obtained values of Q are 0.372, 0.521, and 0.600, respectively, and the community structures found are meaningful.

5.3.2.3 Comparison between MEAs-SN and Existing Algorithms

In this experiment, MEA$_s$-SN is compared with three existing algorithms, namely FEC [50], the Louvain method [53, 54], and CSA$_{HC}$-SN [21]. FEC is an algorithm recently proposed for SNs and obtained a good performance. The Louvain method is a special version for SNs of the algorithm in [42] and is implemented in Pajek [30]. CSA$_{HC}$-SN is a memetic algorithm for SNs proposed in [21] that can optimize both Q_{signed} and the improved modularity density D. In the following experiment, FEC and CSA$_{HC}$-SN run under the same experimental environment with MEA$_s$-SN. Both Q_{signed} and D are optimized by CSA$_{HC}$-SN, labeled as CSA$_{HC}$-SN(Q) and CSA$_{HC}$-SN(D), respectively. The Louvain method runs under Pajek.

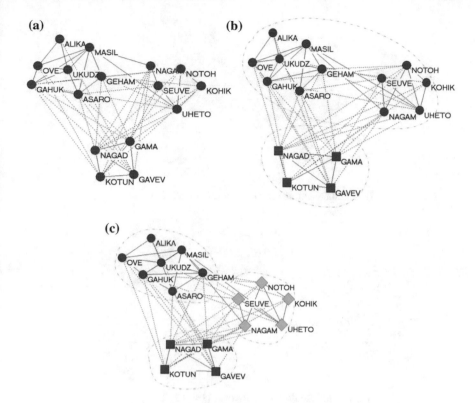

Fig. 5.19 **a** The topological structure of the Gahuku-Gama Subtribes Network. **b** and **c** are the two community structures obtained by MEA$_s$-SN

All these four algorithms are tested on the synthetic SNs with $\mu = 0.1 \sim 0.5$ and $P_+ = 0 \sim 1$ systematically. Since previous results showed that MEA$_s$-SN and FEC are sensitive to P_-, and the performance drops when P_- is larger, we only compare the performances of different algorithms when P_- is in the range of $[0, 0.4]$. Since the computational cost of CSA$_{HC}$-SN is too high for large networks, the network size for CSA$_{HC}$-SN is set to 1,000 and 10,000 for three other algorithms. For each network, 10 independent runs are conducted for each algorithm and the results are shown in Fig. 5.21.

As can be seen, first, MEA$_s$-SN outperforms CSA$_{HC}$-SN(Q) and CSA$_{HC}$-SN(D) obviously almost in all parameter combinations. Some results for $P_+ = 1.0$ of CSA$_{HC}$-SN(Q) and CSA$_{HC}$-SN(D) are missing due to the high computational cost.

For the Louvain method, when $P_- = 0$, the *NMI* is always higher than 0.8 when μ increases from 0.1 to 0.5 and P_+ increases from 0 to 1. The *NMI* of MEA$_s$-SN is also always higher than 0.8 in these parameter combinations, and better than that of the Louvain method in most cases. When P_- increases, the performance of the Louvain method drops dramatically, which is also very sensitive to P_+; that is, when P_+ increases, the performance drops dramatically, too. Although the performance

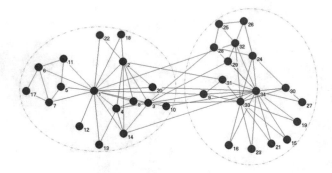

(a) Zachary karate club network, $Q = 0.372$

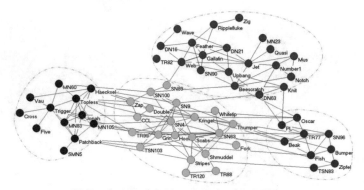

(b) Bottlenose dolphin network, $Q = 0.521$

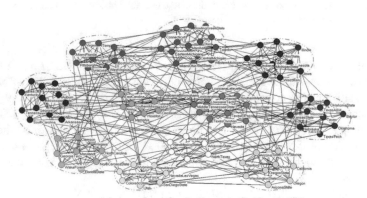

(c) American football network, $Q = 0.600$

Fig. 5.20 The community structures found by MEA$_s$-SN for the three popular unsigned benchmark networks

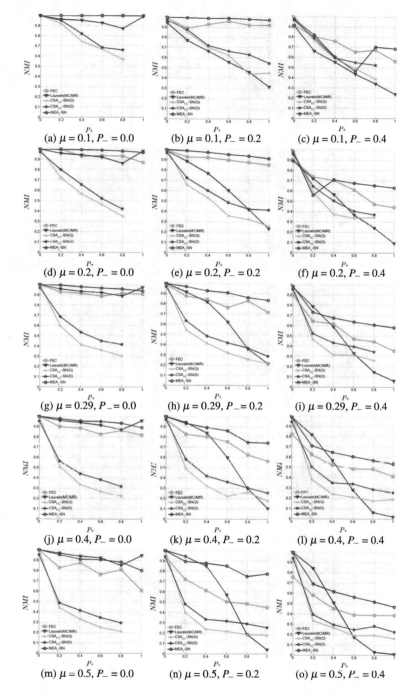

Fig. 5.21 The comparison between MEA$_s$-SN and existing algorithms

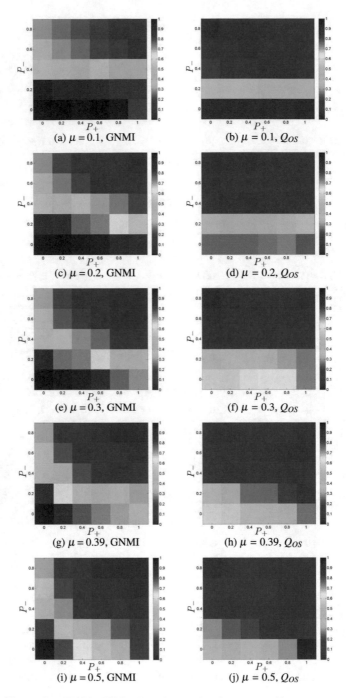

Fig. 5.22 The results of MEA$_s$-SN for detecting overlapping communities

of MEA$_s$-SN also drops, it is clearly better than that of the Louvain method, which illustrates that MEA$_s$-SN is more robust to noises.

For FEC, when $P_- = 0$, the *NMI* is always higher than 0.8 except when $\mu = 0.5$, and MEA$_s$-SN performs similar or better than FEC for these parameter combinations. When $P_- \geq 0.2$, the performance of FEC also drops, but is not so sensitive to the increase of P_+ like Louvain method. To compare with our method, FEC is outperformed by MEA$_s$-SN for all parameter combinations when $P_- = 0.2$. When $P_- = 0.4$, MEA$_s$-SN always outperforms FEC when $\mu \geq 0.3$.

5.3.2.4 Experiments on Overlapping Community Detection From Synthetic SNs

In this experiment, synthetic SNs were used to test the performance of MEA$_s$-SN in detecting overlapping communities. Parameters *on* and *om* were set to 100 and 2, respectively, and the number of nodes was set to 1,000. Other parameters of the generator were the same with those in the previous section. We also systematically tested the effect of μ, P_+, and P_-. In addition, μ increased from 0.1 to 0.5, P_+ increased from 0 to 1, and P_- increased from 0 to 0.8. For each combination of these three parameters, 30 independent runs of MEA$_s$-SN were conducted, and the averaged *NMI* and Q_{signed} are reported in Fig. 5.22.

Compared with detecting separated communities, detecting overlapping communities is more difficult, which is demonstrated by the obtained Q_{os}. We can see that Q_{os} decreases dramatically when μ, P_+, and P_- increase; that is to say, the community structure is getting more and more ambiguous quickly. However, Fig. 5.22 shows that MEA$_s$-SN still obtains a good performance when μ, P_-, and P_+ are not too large. For example, when $\mu \leq 0.3$, $P_+ \leq 0.4$, and $P_- \leq 0.2$, the *NMI* is always larger than 0.8. In fact, even when $0.4 \leq P_+ \leq 0.8$, the *NMI* is still larger than 0.5. For $\mu \geq 0.4$, the community structure gets very ambiguous, and the *NMI* decreases quickly accordingly. However, when the noise levels are not too large; that is, when both P_+ and P_- are smaller than 0.2, the obtained *NMI* is still larger than 0.5.

References

1. Agarwal, G., Kempe, D.: Modularity-maximizing graph communities via mathematical programming. The Eur. Phys. J. B **66**(3), 409–418 (2008)
2. Andrews, G.E.: The Theory of Partitions, Volume 2 of Encyclopedia of Mathematics and its Applications, vol. 19, no. 76, p. 18. Addison-Wesley (1976)
3. Bui, T.N., Moon, B.R.: Genetic algorithm and graph partitioning. IEEE Trans. Comput. **45**(7), 841–855 (1996)
4. Danon, L., Diaz-Guilera, A., Duch, J., Arenas, A.: Comparing community structure identification. J. Stat. Mech. Theory Exp. **2005**(09) (2005). DOI P09008
5. Doreian, P., Mrvar, A.: A partitioning approach to structural balance. Soc. Netw. **18**(2), 149–168 (1996)

6. Easley, D., Kleinberg, J.: Networks, Crowds, and Markets: Reasoning About a Highly Connected World. Cambridge University Press (2010)
7. Fortunato, S., Barthélemy, M.: Resolution limit in community detection. Proc. Natl. Acad. Sci. **104**(1), 36–41 (2007)
8. Fortunato, S., Castellano, C.: Community structure in graphs. In: Computational Complexity, pp. 490–512. Springer (2012)
9. Girvan, M., Newman, M.E.: Community structure in social and biological networks. Proc. Natl. Acad. Sci. **99**(12), 7821–7826 (2002)
10. Gog, A., Dumitrescu, D., Hirsbrunner, B.: Community detection in complex networks using collaborative evolutionary algorithms. In: Advances in Artificial Life, pp. 886–894 (2007)
11. Goldberg, D.E., Lingle, R., et al.: Alleles, loci, and the traveling salesman problem. In: Proceedings of an International Conference on Genetic Algorithms and their Applications, vol. 154, pp. 154–159. Lawrence Erlbaum, Hillsdale, NJ (1985)
12. Gómez, S., Jensen, P., Arenas, A.: Analysis of community structure in networks of correlated data. Phys. Rev. E **80**(1) (2009). DOI 016114
13. Gong, M., Fu, B., Jiao, L., Du, H.: Memetic algorithm for community detection in networks. Phys. Rev. E **84**(5) (2011). DOI 056101
14. Huang, J., Sun, H., Liu, Y., Song, Q., Weninger, T.: Towards online multiresolution community detection in large-scale networks. PloS one **6**(8) (2011). DOI e23829
15. Krebs, V.: Books about us politics (2004). http://www.orgnet.com
16. Kropivnik, S., Mrvar, A.: An analysis of the slovene parliamentary parties network. In: Developments in Data Analysis, pp. 209–216 (1996)
17. Kunegis, J., Preusse, J., Schwagereit, F.: What is the added value of negative links in online social networks? In: Proceedings of the 22nd International Conference on World Wide Web, pp. 727–736. ACM (2013)
18. Lancichinetti, A., Fortunato, S., Kertész, J.: Detecting the overlapping and hierarchical community structure in complex networks. New J. Phys. **11**(3) (2009). DOI 033015
19. Lancichinetti, A., Fortunato, S., Radicchi, F.: Benchmark graphs for testing community detection algorithms. Phys. Rev. E **78**(4) (2008). DOI 046110
20. Leicht, E.A., Newman, M.E.: Community structure in directed networks. Phys. Rev. Lett. **100**(11) (2008). DOI 118703
21. Li, Y., Liu, J., Liu, C.: A comparative analysis of evolutionary and memetic algorithms for community detection from signed social networks. Soft Comput. **18**(2), 329–348 (2014)
22. Li, Z., Liu, J.: A multi-agent genetic algorithm for community detection in complex networks. Phys. A Stat. Mech. Appl. **449**, 336–347 (2016)
23. Liu, C., Liu, J., Jiang, Z.: A multiobjective evolutionary algorithm based on similarity for community detection from signed social networks. IEEE Trans. Cybern. **44**(12), 2274–2287 (2014)
24. Liu, J.: Autonomous Agents and Multi-agent Systems: Explorations in Learning, Self-organization and Adaptive Computation. World Scientific (2001)
25. Liu, J., Jing, H., Tang, Y.Y.: Multi-agent oriented constraint satisfaction. Artif. Intell. **136**(1), 101–144 (2002)
26. Liu, J., Zhong, W., Abbass, H.A., Green, D.G.: Separated and overlapping community detection in complex networks using multiobjective evolutionary algorithms. In: 2010 IEEE Congress on Evolutionary Computation (CEC), pp. 1–7. IEEE (2010)
27. Lovász, L.: Combinatorial Problems and Exercises, vol. 361. American Mathematical Society (1993)
28. Lusseau, D., Schneider, K., Boisseau, O.J., Haase, P., Slooten, E., Dawson, S.M.: The bottlenose dolphin community of doubtful sound features a large proportion of long-lasting associations. Behav. Ecol. Sociobiol. **54**(4), 396–405 (2003)
29. Medus, A., Acuña, G., Dorso, C.: Detection of community structures in networks via global optimization. Phys. A Stat. Mech. Appl. **358**(2), 593–604 (2005)
30. Mrvar, A., Batagelj, V., Pajek Pajek-XXL: Programs for analysis and visualization of very large networks. In: Reference Manual. University of Ljubljana, Ljubljana (2013)

31. Naeni, L.M., Berretta, R., Moscato, P.: Ma-net: a reliable memetic algorithm for community detection by modularity optimization. In: Proceedings of the 18th Asia Pacific Symposium on Intelligent and Evolutionary Systems, vol. 1, pp. 311–323. Springer (2015)
32. Newman, M.E.: Fast algorithm for detecting community structure in networks. Phys. Rev. E **69**(6) (2004). DOI 066133
33. Newman, M.E., Girvan, M.: Finding and evaluating community structure in networks. Phys. Rev. E **69**(2) (2004)
34. Noack, A., Rotta, R.: Multi-level Algorithms for Modularity Clustering, pp. 257–268. Springer (2009)
35. Palla, G., Derényi, I., Farkas, I., Vicsek, T.: Uncovering the overlapping community structure of complex networks in nature and society. Nature **435**, 814–818 (2005)
36. Park, Y., Song, M.: A genetic algorithm for clustering problems. In: Proceedings of the Third Annual Conference on Genetic Programming 1998, pp. 568–575 (1998)
37. Pizzuti, C.: Ga-net: a genetic algorithm for community detection in social networks. In: PPSN, pp. 1081–1090 (2008)
38. Pizzuti, C.: A multiobjective genetic algorithm to find communities in complex networks. IEEE Trans. Evol. Comput. **16**(3), 418–430 (2012)
39. Pólya, G., Szegö, G.: Operations with power series. In: Problems and Theorems in Analysis I, pp. 1–15. Springer (1998)
40. Read, K.E.: Cultures of the central highlands, new guinea. Southwest. J. Anthropol. **10**(1), 1–43 (1954)
41. Rosvall, M., Bergstrom, C.T.: Maps of random walks on complex networks reveal community structure. Proc. Natl. Acad. Sci. **105**(4), 1118–1123 (2008)
42. Rotta, R., Noack, A.: Multilevel local search algorithms for modularity clustering. J. Exp. Algorithm. (JEA) **16**, 2–3 (2011)
43. Shen, H., Cheng, X., Cai, K., Hu, M.B.: Detect overlapping and hierarchical community structure in networks. Phys. A Stat. Mech. Appl. **388**(8), 1706–1712 (2009)
44. Shi, C., Yan, Z., Cai, Y., Wu, B.: Multi-objective community detection in complex networks. Appl. Soft Comput. **12**(2), 850–859 (2012)
45. Shi, C., Yu, P.S., Cai, Y., Yan, Z., Wu, B.: On Selection of Objective Functions in Multi-objective Community Detection, pp. 2301–2304 (2011)
46. Talbi, E.G., Bessiere, P.: A parallel genetic algorithm for the graph partitioning problem. In: Proceedings of the 5th International Conference on Supercomputing, pp. 312–320. ACM (1991)
47. Tang, J., Lou, T., Kleinberg, J.: Inferring social ties across heterogenous networks. In: Proceedings of the Fifth ACM International Conference on Web Search and Data Mining, pp. 743–752. ACM (2012)
48. Tasgin, M., Herdagdelen, A., Bingol, H.: Community detection in complex networks using genetic algorithms (2007)
49. Xu, G., Tsoka, S., Papageorgiou, L.G.: Finding community structures in complex networks using mixed integer optimisation. The Eur. Phys. J. B-Condens. Matter Complex Syst. **60**(2), 231–239 (2007)
50. Yang, B., Cheung, W., Liu, J.: Community mining from signed social networks. IEEE Trans. Knowl. Data Eng. **19**(10) (2007)
51. Ye, Z., Hu, S., Yu, J.: Adaptive clustering algorithm for community detection in complex networks. phys. Rev. E **78**(4) (2008). DOI 046115
52. Zachary, W.W.: An information flow model for conflict and fission in small groups. J. Anthropol. Res. **33**(4), 452–473 (1977)
53. Zhang, Q., Li, H.: Moea/d: a multiobjective evolutionary algorithm based on decomposition. IEEE Trans. Evol. Comput. **11**(6), 712–731 (2007)

Chapter 6
Evolving Robust Networks Using Evolutionary Algorithms

Most interacting systems in nature and society can be modeled as complex networks with the components of systems being represented as nodes and the relationships between components being represented as edges [2, 14, 27]. Network properties, such as small-world property [33] and scale-free property [1, 4], have attracted increasing attention.

Moreover, one extremely important aspect of networks is their capability to withstand failures and fluctuations in the functionality of their nodes and links, their robustness. In reality, failures may occur in many different ways and to a different degree, depending on the complexity of the system under examination. Thus, the robustness of networks is of great importance to guarantee the security of network systems, such as those in airports, the power grids, the transportation, the World Wide Web, and the disease control networks. The breakdown of many real-world networks caused by the failure on a small number of nodes or links has led to considerable economic losses in the past, such as the breakdown of the Chinese power line network caused by the bad weather [38], and the effects of the subprime mortgage crisis in the USA on the worldwide economic network [37]. Therefore, the robustness of different network structures has been studied intensively in the past decade [15, 31, 34, 37].

6.1 Network Robustness and Analysis

Usually, a network is robust if their function is not severely affected by the attacks to nodes or links, which can be either random or malicious. To study the network robustness, suitable measures are needed to evaluate the robustness. Recently, many robustness measures have been proposed from different aspects. One important factor in designing measures is attacks. In random attacks [3], each node or edge is removed with the same probability. While in malicious attacks [12], one popular

© Springer Nature Switzerland AG 2019
J. Liu et al., *Evolutionary Computation and Complex Networks*,
https://doi.org/10.1007/978-3-319-60000-0_6

way is to sequentially remove the most important node or edge at each attack, and the importance of remaining nodes or edges is re-calculated. This process repeats until only isolated nodes are left in the network. For evaluating the robustness of networks and guiding the robustness optimization process, several robustness measures have been proposed, and we give a brief review here.

6.1.1   Robustness Measures Based on Connectivity

In 1970, Frank et al. [18] analyzed the survivability of networks using the basic concept of graph connectivity. However, the graph connectivity only partly reflects the ability of graphs to retain certain degree of connectedness under deletion. Thus, other improved measures were introduced, such as superconnectivity [5] and conditional connectivity [20]. Although these improved measures consider both the cost of damaging a network and the extent to which the network is damaged, the computational cost in calculating these measures for general graphs is too high and these measures are hard to apply in related studies.

The edge connectivity, labeled as $\upsilon(G)$, is the minimum number of edges that must be removed from a connected graph in order to disconnect it. Let $\upsilon_{s-t}(G)$ be the number of edges that must be removed to disconnect nodes s and t, then the edge connectivity is defined as

$$\upsilon(G) = \min_{s,t\neq s\in V} \{\upsilon_{s-t}(G)\} \tag{6.1}$$

According the above definition, it is not difficult to find that the larger the value of $\upsilon(G)$ is, the more robust the network is, because more edges need to be removed to disconnect networks. But the value is not larger than the minimum node degree in the network. So the value of this measure cannot be changed too much through adjusting the network topology without changing the degree distribution and each nodal degree. Similarly, the node connectivity, being similar with the edge connectivity, is the minimum number of nodes that must be removed from a connected graph in order to disconnect it. Let $\omega(G)$ be the node connectivity of network G, and $\omega_{s-t}(G)$ be the minimum number of nodes that must be removed to disconnect nodes s and t. Thus,

$$\omega(G) = \min_{s,t\neq s\in V \wedge e_{st}\notin E} \{\omega_{st}(G)\} \tag{6.2}$$

According the definition, $\omega(G)$ is not defined in a full connected graph. The larger the value of $\omega(G)$ is, the more robust the graph is, because more nodes need to be removed to disconnect the graph. A remarkable result of these two robustness measures is that $\omega_{s-t}(G) \leq \upsilon_{s-t}(G)$ for any s and t ($s \neq t \in V$, $e_{st} \notin E$), so $\omega(G) \leq \upsilon(G)$ [6]. Thus, the value of this measure cannot be changed too much through adjusting the topology without changing the degree distribution and each nodal degree.

6.1.2   Robustness Measures Based on Random Graph Theory

Except for evaluating the robustness based on graph connectivity, the random graph theory is another viewpoint from which to define measures, which is more popular in recent studies. In 2000, Albert et al. [3] studied the network robustness through random graph theory. They proposed a statistical measure, namely the critical removal fraction of vertices (edges) for the disintegration of networks, to characterize the structural robustness of complex networks. The disintegration of networks is measured in terms of network performance. The most common performance measurements include the diameter, the size of the largest component, the average path length, and communication efficiency [23]. The critical fraction against random attacks is labeled as p_c^r. According to [29], p_c^r for any degree distribution $P(k)$ is calculated as

$$p_c^r = 1 - \frac{1}{\kappa_0 - 1} \tag{6.3}$$

where κ_0 is equal to $\langle \kappa \rangle / \langle \kappa^2 \rangle$, $\langle \kappa \rangle$ is the average nodal degree of the original network, and $\langle \kappa^2 \rangle$ is the average of square of nodal degree. The larger the value of p_c^r is, the more robust the network is. If the degree distribution and each nodal degree are unchanged, then κ_0 is unchanged.

The critical fraction against targeted attacks is labeled as p_c^t. In [11], p_c^t is defined against malicious attacks based on nodal degree. After the node with the largest degree is removed, the degree distribution needs to be re-calculated. In fact, we can calculate p_c^t for graphs with any degree distributions through simulating the attacking process. Attack the node with the largest degree, check the condition of networks, and repeat this process until the left network is disconnected, then p_c^t is obtained. The larger the value of is, the more robust the network is.

6.1.3   Robustness Measures R and Its Extensions

In 2011, Schneider et al. [32] pointed out that the robustness measure in terms of the critical fraction of attacks at which the system completely collapse, the percolation threshold, may not be useful in many realistic cases. These measures, for example, ignore the situations in which the network suffers from a significant damage, but still keeps its integrity. Thus, they proposed a unique robustness measure R, which considers the size of the largest component against malicious node attacks,

$$R = \frac{1}{N} \sum_{Q=1}^{N} s(Q) \tag{6.4}$$

where $s(Q)$ is the fraction of nodes of the largest connected component after removing Q nodes. The normalization factor $1/N$ ensures that the robustness of networks with different sizes can be compared. The larger the value of R is, the more robust the network is.

Since the original R only considers the attacks on nodes, Zeng et al. extended it to consider the attacks on links in [36] as follows:

$$R_l = \frac{1}{M} \sum_{P=1}^{M} s(P) \tag{6.5}$$

where $s(P)$ is the fraction of nodes in the largest connected component after removing P links. The normalization factor $1/M$ also ensures that the robustness of network with different sizes can be compared. The larger the value of R_l is, the more robust the network is.

Instead of keeping track of the size of largest connected subgraph, Louzada et al. in [25] focused on the communication efficiency [23] of networks after each attack and proposed a new measure, namely integral efficiency of network, which is labeled as $IntE$,

$$IntE = \frac{1}{N} \sum_{Q=1}^{N} E(Q) \tag{6.6}$$

where $E(Q)$ is the communication efficiency of networks after removing Q nodes. The normalization factor $1/N$ also ensures that the robustness of networks with different sizes can be compared. The larger the value of $IntE$ is, the more robust the network is. Furthermore, $E(0)$ is the communication efficiency of original networks, which can be calculated as

$$E(G) = \frac{1}{N(N-1)} \sum_{i \neq j \in G} \zeta_{ij} = \frac{1}{N(N-1)} \sum_{i \neq j \in G} \frac{1}{d_{ij}} \tag{6.7}$$

where ζ_{ij} is the communication efficiency between vertices i and j which is defined to be inversely proportional to the shortest distance: $\zeta_{ij} = d_{ij}$. When there is no path between i and j, $d_{ij} = +\infty$ and $\zeta_{ij} = 0$.

Combining R_l with $IntE$, the communication efficiency can be also tracked for malicious link attacks, which extends the measure as follows:

$$IntE_l = \frac{1}{M} \sum_{P=1}^{M} E(P) \tag{6.8}$$

where $E(P)$ is the communication efficiency of network, after removing P edges, and is calculated through (7.6). The normalization factor $1/M$ also ensures that the robustness of networks with different sizes can be compared. The larger the value of $IntE_l$ is, the more robust the network is.

6.1.4 Robustness Measures Based on Eigenvalue

A remarkable group of robustness measures are those based on the eigenvalue of the network Laplacian matrix or adjacent matrix. Here, two robustness measures, namely algebraic connectivity, which is based on the eigenvalue of Laplacian matrix, and natural connectivity, which is based on the eigenvalue of adjacent matrix, are introduced in detail.

The algebraic connectivity, $a(G)$, is the second smallest eigenvalue of the Laplacian matrix. Fiedler [17] showed that the magnitude of the algebraic connectivity reflects how well connected the overall graph is.

$$\alpha(G) = \lambda_i, \quad \lambda_1 \le \lambda_2 \le \lambda_3 \le \cdots \le \lambda_N \tag{6.9}$$

where $\lambda_i, i = 1, 2, 3 \ldots, N$ are the eigenvalue of Laplacian matrix of graph G.

However, the algebraic connectivity is too coarse to capture important features of structural robustness of complex networks [34]. Thus, Wu et al. [34] proposed the natural connectivity, which characterizes the redundancy of alternative routes in a network by quantifying the weighed number of closed walks of all lengths. The natural connectivity can be regarded as an average eigenvalue that changes strictly monotonically with the addition or deletion of edges.

$$\bar{\lambda} = \ln \left(\frac{1}{N} \sum_{i=1}^{N} e^{\lambda_i} \right) \tag{6.10}$$

where λ_i is the ith eigenvalue of the graph adjacency matrix. Given the number of vertices, the empty graph has the minimum *natural connectivity*, and the complete graph has the maximum *natural connectivity*. So the lager the value of λ is, the more robust the network is.

6.1.5 Comparison Among These Measures

One of the major functions of these robustness measures is to guide the optimization process as an optimization objective to find more robust networks. This section focuses on studying of the performance of different measures when optimizing network robustness guided separately by the above six types of measures, $\upsilon(G)$, $\omega(G)$, p_c^r, p_c^t, R, R_l, $IntE$, $a(G)$, and to see whether the networks which are robust in terms of certain measures are robust in terms of other measures. In the optimization process, we conduct hill-climbing algorithm as in [32] to improve the robustness of corresponding networks. The results are given in Fig. 6.1.

From Fig. 6.1, we can see that the optimized networks are not robust in every aspect. Figure 6.1a, b, f, and h shows that these optimized networks are not robust in terms of $\bar{\lambda}$. p_r^c only depends on the degree distribution, so the optimization without

changing the degree distribution cannot affect it. Figure 6.1d, e, and g shows almost
similar phenomena, because all of them are defined on the basis of malicious nodal
attack. In terms of network sizes, networks with different numbers of nodes show
similar performances in the optimization processes. Moreover, we show the change of
average shortest path in the process of robustness optimization guided by different
measures in Fig. 6.2, to depict the variation of network properties in the rewiring
process.

Figure 6.2 shows when R and $\bar{\lambda}$ increase, the average shortest path length increases
dramatically. But for the four other measures, the average shortest path length changes
just slightly. Thus, in the process of optimizing R and $\bar{\lambda}$, the shortest path is sacrificed.
In the process of optimizing $IntE$, the average shortest path length first increases
and then decreases, and during the process of optimizing R_l, it is almost not changed.

From the above results, we can see a robust network in terms of certain robustness
measure may be fragile in terms of other measures, especially when these measures
are uncorrelated or negatively correlated with each other. However, if the robustness
measures are closely correlated with each other, they can often be improved together.
Furthermore, the results indicate that it is not simple to apply a robust network

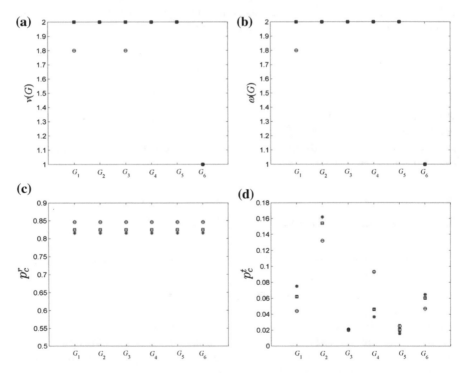

Fig. 6.1 The values of robustness measures of different optimized networks. The labels $G_1 \sim G_6$
denote p_c^t, R, R_l, $IntE$, $a(G)$, and $lambda$, respectively. In this figure, circles represent the results
of networks with 100 nodes, squares represent those with 300 nodes, and asterisks represent those
with 500 nodes

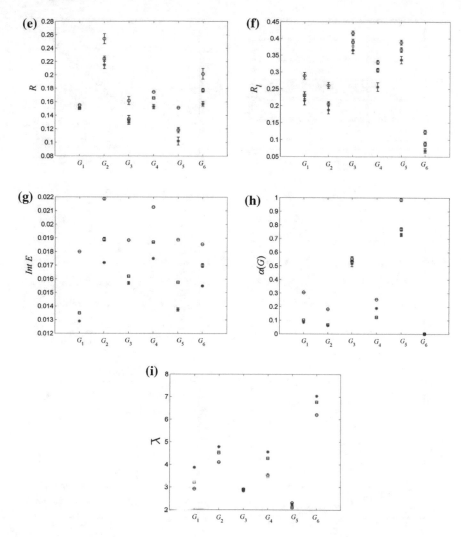

Fig. 6.1 (continued)

into practical situations, and there are many network characteristics that need to be considered and balanced. During the optimization process, taking more suitable robustness measures into consideration is a promising way to generate more robust networks.

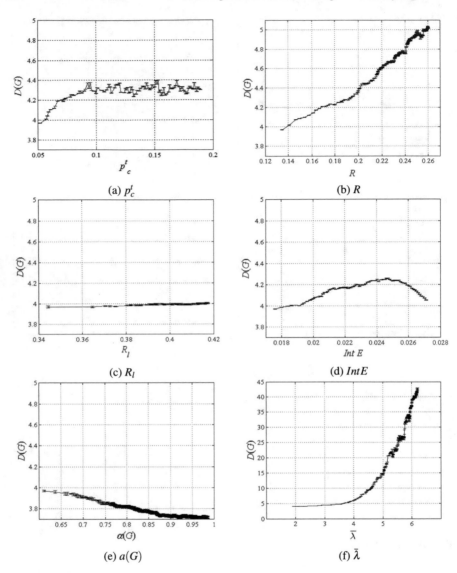

Fig. 6.2 The relationships between the average shortest path length and the robustness measures of networks obtained during the optimization process

6.2 Evolutionary Algorithms for Networks Robustness Optimization

In reality, it is significant to improve the robustness of networked systems against potential attacks. Generally, scale-free (SF) networks exist broadly in modern applications, so we mainly consider the enhancement of SF networks in this section.

In terms of network property, the fragileness of SF networks under the malicious attacks comes from their heavy-tailed property, causing loss of a large number of links when a hub node is crashed. The heavy loss of network links quickly makes the network to be sparsely connected and then fragmented. Thus, the major purpose of this work is to study how to improve the robustness of SF networks against malicious attacks. A simple solution for this problem is to add links, but additional links also increase the costs significantly. Therefore, we consider how to improve the network robustness against malicious attacks based on nodal degree without changing the degree distribution of the initial networks.

6.2.1 Introduction of MA-RSF$_{MA}$

A few studies have been proposed to tackle this problem. Xiao et al. in [35] designed a simple rewiring method that does not change any nodal degrees, and they showed that network robustness can be steadily enhanced at a slightly decreased assortativity coefficient. In [32], Schneider et al. introduced a new measure for robustness, which considers the size of the largest connected cluster during the entire attack process, and used this measure to devise a heuristic method to mitigate the malicious attacks. Based on the same measure, Buesser et al. in [10] proposed a simulated annealing algorithm, and Louzada et al. in [24] proposed a smart rewiring method for this problem. All these methods manifest a good performance in improving the network robustness over the initial networks. However, these methods can still be improved by considering the possibility of global searches in the optimization algorithm. Thus, more powerful methods that can conduct both global and local searches are needed to optimize the network structure.

Evolutionary algorithms (EAs), which are a kind of stochastic global optimization methods inspired by the biological mechanism of evolution and heredity, have been successfully used to solve various hard optimization problems. One popular branch in the field of EAs is memetic algorithms (MAs), where the concept of "memetic" came from Dawkin's concept of a meme, which represents a unit of cultural evolution that can exhibit local refinement [26, 30]. In MAs, a meme is generally considered as an individual learning procedure capable of performing local refinements. Thus, MAs successfully combine global and local searches, and have been shown to be more efficient and more effective than traditional EAs for many problems [22, 28]. In [37], Zhou et al. proposed a memetic algorithm, named as MA-RSF$_{MA}$, to enhance the robustness guided by R of SF networks against malicious nodal attacks, which has shown effectiveness in optimization, and we give a short review here.

In MA-RSF$_{MA}$, each chromosome represents a graph, thus, a population with Ω chromosomes represented Ω graphs, which are labeled as $G_1, G_2, \ldots, G_\Omega$. In the initialization, since we need to keep invariant the number of links and the degree of each node, each chromosome is generated by randomly adjusting a fraction of edges in the initial scale-free network G_0. To adjust the edges in G_0, the swap operation proposed in [32] is employed; that is, the connections of two randomly chosen edges which have no common nodes are swapped, and any swap operations that can keep the graph being connected are accepted without checking whether the robustness of the network is improved or not. This initialization process is summarized in Algorithm 6.1.

Algorithm 6.1 Population Initialization

Input:
  Ω: Population size;
  G_0: Initial scale-free network;
Output:
  $\mathbf{P}^1 = \{G_1^1, G_2^1, \ldots, G_\Omega^1\}$: Population for the 1st generation;

---

for $i = 1$ **to** Ω **do**
  $G_i^1 \leftarrow G_0$;
  for $j = 1$ **to** M **do**
    Randomly select two edges e_{kl} and e_{mn} from G_i^1, where m, n are different from k, l and e_{km} and e_{ln} do not exist in G_i^1;
    Remove e_{kl} and e_{mn} from G_i^1, and add e_{km} and e_{ln} to G_i^1;
    if G_i^1 is not connected **then**
      Remove e_{km} and e_{ln} from G_i^1, and add e_{kl} and e_{mn} back to G_i^1;
    end if
  end for
end for

The swap operation proposed in [32] is equivalent to a mutation operator in memetic algorithm and can only search in small area. Another effective operator in memetic algorithm is the crossover operator, that can search in a large area. A new crossover operator which works on two parent chromosomes and can keep invariant the degree of each node in the child chromosomes is implemented in MA-RSF$_{MA}$. The details are as follows.

Suppose G_{p1} and G_{p2} are two parent chromosomes, and G_{c1} and G_{c2} are two child chromosomes. First assign G_{p1} to G_{c1} and G_{p2} to G_{c2}, then for each node i in G_{c1} and G_{c2} do the following operations with the probability p_c. Obtain the following sets of nodes,

$$V_i^{G_{c1}} = \{j \mid e_{ij} \in E^{G_{c1}}\} \tag{6.11}$$

$$V_i^{G_{c2}} = \{j \mid e_{ij} \in E^{G_{c2}}\} \tag{6.12}$$

$$\bar{V}_i^{G_{c1}} = V_i^{G_{c1}} - (V_i^{G_{c1}} \mid V_i^{G_{c2}}) \tag{6.13}$$

$$\bar{V}_i^{G_{c2}} = V_i^{G_{c2}} - (V_i^{G_{c1}} \mid V_i^{G_{c2}}) \tag{6.14}$$

Next, for each node $j \in \bar{V}_i^{G_{c1}}$, select a node that have not been used. Remove e_{ij} from G_{c1} and e_{ik} from G_{c2}. Add e_{ik} to G_{c1} and e_{ij} to G_{c2}. To keep the degree of each node from being invariant, randomly select another edge e_{kl} that node k connects in G_{c1}, then remove e_{kl} and add e_{jl}. The similar operations are also needed to be conducted on G_{c2}. In this way, a part of network structure of G_{p1} is transformed into G_{p2}, and vice versa. The details of this crossover operator are summarized in Algorithm 6.2

Algorithm 6.2 Crossover Operator

Input:
  G_{p1} and G_{p2}: Two parent chromosomes;
  p_c: Crossover rate;
Output:
  G_{c1} and G_{c2}: Two child chromosomes;

$G_{c1} \leftarrow G_{p1}, G_{c2} \leftarrow G_{p2}$;
for $i = 1$ **to** N **do**
  if $U(0, 1) < p_c$ **then** /*$U(0, 1)$ *is a uniformly distributed random real number in* [0, 1];*/
    Determine $\overline{V}_i^{G_{c1}}$ and $\overline{V}_i^{G_{c2}}$;
    for each node $j \in \overline{V}_i^{G_{c1}}$ **do**
      Randomly select a node $k \in \overline{V}_i^{G_{c2}}$;
      Remove e_{ij} from G_{c1} and e_{ik} from G_{c2};
      Add e_{ik} to G_{c1} and e_{ij} to G_{c2};
      Randomly select another edge e_{kl} that node k connects in G_{c1} and e_{jl} does not exist in G_{c1};
      Remove e_{kl} and add e_{jl} in G_{c1};
      Randomly select another edge e_{jm} that node j connects in G_{c2} and e_{km} does not exist in G_{c2};
      Remove e_{jm} and add e_{km} in G_{c2};
      $\overline{V}_i^{G_{c2}} = \overline{V}_i^{G_{c2}} - \{k\}$;
    end for
  end if
end for

The local search operator is another kind of important search operations in memetic algorithms. Considering the conclusion drawn by Schneider et al. in [21, 32] drew the conclusion that SF networks with an "onion structure" are very robust against targeted high-degree attacks. The onion structure is a network where nodes with almost the same degree are connected. In MA-RSF$_{\text{MA}}$, the local search operator is designed to let the nodes prefer to connect to others with similar degree. Suppose edges e_{ij} and e_{kl} are selected to swap, if

$$| d_i - d_k | + | d_j - d_l | < \alpha \times \left(| d_i - d_j | + | d_k - d_l | \right) \qquad (6.15)$$

then e_{ij} and e_{kl} are replaced by e_{ik} and e_{jl}, where d_i, d_k, d_j, and d_l are the degrees of corresponding nodes, and α is a parameter that is in the range of [2, 13]. As can be seen, only when the difference between the degrees of the two pairs of nodes is smaller than certain percent of the previous difference is the swap is accepted. Thus, the percent of decreased difference is controlled by α; that is, the larger the value of α is, the weaker the condition is, and the smaller the difference is decreased. In [37], Zhou et al. found that MA-RSF$_{\text{MA}}$ obtains the best search results when α is set to be 0.9. The details of this operator are summarized in Algorithm 6.3.

Algorithm 6.3 Local Search Operator

Input:
   G One chromosome;
   p_l: Local search probability;
   α: Predefined parameter;
Output:
   G: Chromosome after performs the local search operator;

---

for each existing edge e_{ij} **do**
   if $U(0, 1) < p_l$ **then** /*$U(0, 1)$ *is a uniformly distributed random real number in* [0, 1];*/
      Randomly select another existing edge e_{kl};
      if $|d_i - d_k| + |d_j - d_l| < \alpha \times (|d_i - d_j| + |d_k - d_l|)$ **then**
         $G^* \leftarrow G$;
         Remove e_{ij} and e_{kl} from G^*;
         Add e_{ik} and e_{jl} to G^*;
         if $R(G^*) > R(G)$ **then**
            $G \leftarrow G^*$;
         end if
      end if
   end if
end for

In each generation of MA-RSF$_{MA}$, the crossover operator is performed on the population first, then the local search operator is conducted on Ω chromosomes with repetition which are selected out based on their robustness. The framework of MA-RSF$_{MA}$ is summarized in Algorithm 6.4.

Algorithm 6.4 MA-RSF$_{MA}$

Input:
  G_0 Initial scale-free network;
  Ω: Population size;
  p_c: Crossover rate:
  p_l: Local search probability;
  α: Predefined parameter for the local search operator;
Output:
  G^*: Chromosome with the highest robustness found;

$\mathbf{P}^1 \leftarrow Population\_Initialization(\Omega, G_0)$;
$t \leftarrow 1$;
while termination criteria are not satisfied **do**
  $\mathbf{P}_c^t \leftarrow \varnothing$; /*$\mathbf{P}_c^t$ is the child population of $\mathbf{P}^t$;*/
  repeat
    Randomly choose two chromosomes G_i^t and G_j^t that have not been selected;
    $(G_{ci}^t, G_{cj}^t) \leftarrow Crossover\_Operator(G_i^t, G_j^t, p_c)$;
    $\mathbf{P}_c^t \leftarrow \mathbf{P}_c^t \bigcup \left\{ G_{ci}^t, G_{cj}^t \right\}$;
  until all chromosomes in $\mathbf{P}^t$ have been selected
  Calculate the robustness of each chromosome in $\mathbf{P}_c^t$;
  for $i = 1$ **to** Ω **do**
    Select a chromosome G^t from $\mathbf{P}^t$ and $\mathbf{P}_c^t$ using the roulette wheel selection based on the robustness of all chromosomes;
    /*In the roulette wheel selection, the probability that a chromosome is selected is proportional to its value of robustness, namely the value of the fitness function.*/
    $G^t \leftarrow Local\_Search\_Operator(G^t, p_l, \alpha)$;
  end for
  $\mathbf{P}^{t+1} \leftarrow$ 2-Tournament_Selection($\mathbf{P}^t, \mathbf{P}_c^t$);
  $t \leftarrow t + 1$;
end while

6.2.2　Experimental Results of MA-RSF$_{MA}$ and Discussions

After BA networks with different scales and edge densities are used to test the performance of MA-RSF$_{MA}$, the optimization results are compared with the following algorithms: hill-climbing algorithm [32], simulated annealing algorithm [10], and smart rewiring algorithm [24]. The numerical results are reported in Fig. 6.3.

As shown in Fig. 6.3, the robustness obtained by MA-RSF$_{MA}$ is always better than that of other algorithms, which illustrates the good performance of memetic

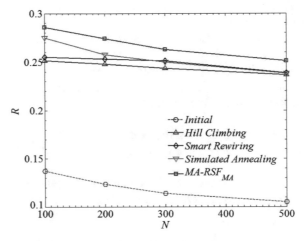

Fig. 6.3 The comparison between MA-RSF$_{MA}$ and existing algorithms on BA networks with different sizes. N is the size of networks

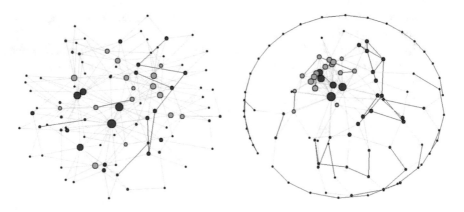

(a) 100 nodes, before optimized, $R = 0.140495$　　(b) 100 nodes, after optimized, $R = 0.280990$

Fig. 6.4 The network topology before and after optimized by MA-RSF$_{MA}$. The size of node is proportional to the node degree, and edges connecting nodes with the same degree are highlighted

algorithm in the task of enhancing network robustness. Intuitively, the impact of robustness optimization on network structure is shown in Fig. 6.4.

In reality, networks are faced with different attacks, including nodes and links can become the punching bag from the view of destructors. Moreover, it is hard to predict the attack target before the destruction happens. Therefore, optimized networks considering multifarious attacking targets have broader applying potentiality (an example will be given in the following section).

6.3 Multi-objective Evolutionary Algorithms for Network Robustness Optimization

In [36], Zeng et al. pointed out that networks which are robust against attacks on nodes may not be robust against attacks on links. On the other hand, in real-world situations, different types of attacks may happen simultaneously. Thus, algorithms which can enhance the robustness of networks against multiple malicious attacks are required. The multi-optimization on different robustness measures can be modeled as a multi-objective optimization problem (MOP). Focusing on solving the MOP of enhancing robustness against different types of attacks, Zhou et al. proposed a two-phase multi-objective EA, termed as MOEA-RSF$_{MMA}$ in [38]. In this section, we first introduce the correlations among some robustness measures, and then we give a brief review of MOEA-RSF$_{MMA}$.

6.3.1 Robustness Measures Correlations

Malicious attacks aim at the most important part of networks, so it is necessary to define the importance of network nodes and links. There are many measures to evaluate the importance of nodes or links, such as degree, betweenness centrality [9, 16], eigenvector centrality [7], and closeness centrality [8, 19]. In these measures, degree and betweenness centrality are more popular, which correspond to the local and global information in a network, respectively.

In terms of degree, there are node degree and edge degree. The node degree is equal to the number of links a node connects to. The edge degree is extended from the node degree [29]. Suppose k_{ij} is the edge degree of link e_{ij}, then

$$k_{ij} = \sqrt{k_i \times k_j} \tag{6.16}$$

where k_i and k_j are the degree of nodes i and j. Usually, the larger the value of degree is, the more important a node or a link is.

In terms of betweenness centrality, there are also node betweenness centrality and edge betweenness centrality. The node betweenness centrality c_l^{BC} of node l is calculated as follows [9]:

$$c_l^{BC} = \sum_{i \neq j \in V} \frac{P_{ij}^l}{P_{ij}} \tag{6.17}$$

where P_{ij} represents the number of shortest paths between nodes i and j and represents the number of shortest paths between nodes i and j that pass through node l.

The edge betweenness centrality $c_{e_{kl}}^{BC}$ of link e_{kl} is defined similarly [9],

$$c_{e_{kl}}^{BC} = \sum_{i \neq j \in V} \frac{P_{ij}^{e_{kl}}}{P_{ij}} \tag{6.18}$$

where $P_{ij}^{e_{kl}}$ represents the number of shortest paths between nodes i and j that pass through e_{kl}. Being similar to degree, the larger the value of betweenness is, the more important a node or a link is.

In [38], Zhou et al. studied the two types of node attacks and two types of link attacks to depict the correlation between different robustness measures. The two types of node attacks are defined as removing the nodes with the highest node degree or the largest node betweenness centrality. All nodes are sorted decreasingly in terms of the node degree or the node betweenness centrality, and then the node with the highest degree or the largest betweenness centrality is removed. After removing a node, the degree or the betweenness centrality of remained nodes is re-calculated and all left nodes are sorted again. The process continues with removing the first node of the sorted list until the network is crashed down. The two types of link attacks are also based on degree and betweenness centrality, similar with node attacks.

In Fig. 6.5, the network robustness against nodal attack based on degree is labeled as R_n^D, based on betweenness centrality is labeled as R_n^{BC}. The network robustness against the link attack based on degree is labeled as R_l^D, based on link betweenness centrality is labeled as R_l^{BC}. The numerical values of Pearson's correlation coefficients between each pairs of these measures are reported in Fig. 6.5.

As can be seen, the Pearson correlation coefficient between R_n^D and R_l^{BC} is the smallest. From the viewpoint of optimization, if two measures strongly positively correlate with each other, they can be combined into one objective, and single-objective algorithms can be used to optimize this combined objective to improve the network structure. However, if two measures strongly negatively correlate with each other, the multi-objective optimization algorithm is a better choice. The conflict between R_n^D and R_l^{BC} is strong, and they may lead the optimization process to find totally different network structures when optimized separately. That is to say, the networks that are robust against malicious nodal attacks based on degree may be vulnerable against malicious link attacks based on betweenness centrality, and vice versa.

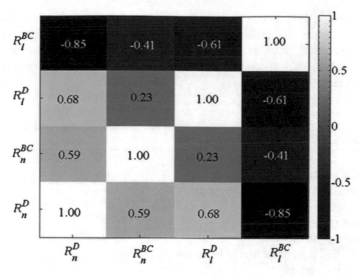

Fig. 6.5 Pearson's correlation coefficients between each pair of robustness measures

6.3.2 Introduction of MOEA-RSF$_{MMA}$

In [38], MOEA-RSF$_{MMA}$ is designed to optimize the robustness of SFNs against malicious node and link attacks without changing the degree distribution and the connectivity of each single node. The multi-optimization problem was modeled as an MOP, and R_n^D and R_l^{BC} were taken as objectives. A brief introduction of MOEA-RSF$_{MMA}$ is given in this section.

Considering the computational costs of evaluating these two objectives are far from equal, MOEA-RSF$_{MMA}$ is composed of two phases, the "R_n^D-sampling" phase and the "R_n^D-R_l^{BC}-optimization" phase, so as to accelerate convergence and improve searching efficiency. The purpose of the R_n^D-sampling phase is to generate a population evenly distributed in the R_n^D-space; that is, generate a set of individuals whose values of R_n^D varies from low to high. Because the above analysis shows that R_n^D and R_l^{BC} are contradicted with each other, the individuals with high values of have low values of R_n^D, and vice versa. Thus, such a population can provide a good starting point for optimizing both R_n^D and R_l^{BC}, and also save the computational cost for calculating R_l^{BC}. Being similar to existing MOEAs, the purpose of the "R_n^D-R_l^{BC}-optimization" phase is to optimize the two objectives together.

In the population, each chromosome represents a graph, thus, a population with Ω chromosomes represents Ω graphs, which are labeled as $G_1, G_2, \ldots, G_\Omega$. Since we need to keep invariant the number of links and the degree of each node, each chromosome can be generated by randomly adjusting a fraction of edges in the initial SFN G_0. Thus, the purpose of the R_n^D-sampling phase is to generate Ω graphs with both high and low R_n^D from the initial graph G_0. However, if we only randomly adjust a fraction of edges in G_0 to generate these Ω graphs, their values of R_n^D will be low. MA-RSF$_{\text{MA}}$ [37], which we introduced in the above section, was employed to realize the purpose of sampling individuals with both high and low R_n^D. The details of the R_n^D-sampling phase are given in Algorithm 6.5, where the operations *Population_Initialization*(.), *Crossover_Operator*(.), and *Local_Search_Operator*(.) are related operations in MA-RSF$_{\text{MA}}$.

Algorithm 6.5 R_n^D-sampling phase

Input:

G_0 Initial scale-free network;

Ω': Population size for MA-RSF$_{\text{MA}}$;

Ω: Population size;

p_c: Crossover rate:

p_l: Local search probability;

α: Predefined parameter for the local search operator;

τ: Sampling interval;

Output:

$\mathbf{P}^1$: Initial population for the "R_n^D-R_l^{BC}-optimization" phase;

$\mathbf{S}^1 \leftarrow$ **Population_Initialization**$(G_o, \Omega'), t \leftarrow 1, \mathbf{P}^1 \leftarrow \varnothing$;

while termination criteria are not satisfied **do**

  $\mathbf{S}_c^t \leftarrow \varnothing$ /*$\mathbf{S}_c^t$ is the child population*/

  repeat

    Randomly choose two chromosomes G_i^t and G_j^t that have not been selected;

    $(G_{ci}^t, G_{cj}^t) \leftarrow$ **Crossover_Operator**(G_i^t, G_j^t, p_c);

    $\mathbf{S}_c^t \leftarrow \mathbf{S}_c^t \bigcup \left\{ G_{ci}^t, G_{cj}^t \right\}$;

  until all chromosomes in $\mathbf{S}^t$ have been selected

  Calculate the value of R_n^D of each chromosome in $\mathbf{S}_c^t$;

  for $i = 1$ **to** Ω' **do**

    Select a chromosome G^t from $\mathbf{S}^t$ and $\mathbf{S}_c^t$ using the roulette wheel selection based on the values of R_n^D of all chromosomes;

    $G^t \leftarrow$ **Local_Search_Operator**(G^t, p_l, α);

  end for

  $\mathbf{S}^{t+1} \leftarrow$ **2-Tournament_selection**$(\mathbf{S}^t, \mathbf{S}_c^t)$;

  if τ fitness function evaluations have been done since the last sampling **then**

    $G' \leftarrow$ The chromosome whith the highest R_n^D in $\mathbf{S}^{t+1}$;

    $P_1 \leftarrow P_1 \bigcup \left\{ G' \right\}$;

  end if

  $t \leftarrow t + 1$;

end while

In the R_n^D-R_l^{BC}-optimization phase, the framework of the well-known multi-objective algorithm NSGA-II [13] is employed. The fast nondominated sort and the crowding distance assignment proposed in [13] are used to generate the next generation. In addition, t crossover operator in [37] is also used. In MOEA-RSF$_{MMA}$, the R_n^D-sampling phase is first conducted so that a good initial population is generated. Then the R_n^D-R_l^{BC}-optimization phase takes over the evolutionary process and further evolves the initial population in terms of both R_n^D and R_l^{BC}. The details of MOEA-RSF$_{MMA}$ are summarized in Algorithm 6.6.

6.3.3 Experimental Results of MOEA-RSF$_{MMA}$ and Discussions

In [38], Zhou et al. tested the performance of MOEA-RSF$_{MMA}$ on different size of SF networks, and they set the parameters of the algorithm as: p_c, p_l, and α are set to 0.5, 0.8, and 0.9. The parameter τ is determined by the number of fitness function evaluations (NFFE) in the R_n^D-sampling phase and the population size in the R_n^D-R_l^{BC}-optimization phase; that is, τ is set to $NFFE/\Omega$. The Pareto fronts obtained by MOEA-RSF$_{MMA}$ are given in Fig. 6.6.

The obtained Pareto fronts in Fig. 6.6 provide a set of networks with different topologies. In general, the networks located at the left part are more robust against link attacks, while those located at the right part are more robust against node attacks, and those in the middle compromise these two types of attacks. Thus, to analyze the topology difference in different parts of the Pareto fronts, three networks were extracted in [38] from each Pareto front, namely the one with the largest R_l^{BC} (the leftmost one), the one with the largest R_n^D (the rightmost one), and the one closest to the coordinate $\left(\frac{\min(R_n^D) + \max(R_n^D)}{2}, \frac{\min(R_l^{BC}) + \max(R_l^{BC})}{2} \right)$ (the middle one), where $\min(R_n^D)$ and $\max(R_n^D)$ are the minimum and maximum values of R_n^D in each Pareto front, and $\min(R_l^{BC})$ and $\max(R_l^{BC})$ are the minimum and maximum value of R_l^{BC} in each Pareto front. These three networks are labeled as G_l, G_r, and G_m, respectively. The topologies of these networks with 100 nodes are shown in Fig. 6.7

As it can be seen in [38], G_r has "onion-like" structures that are consistent with the conclusion drawn in [32]. On the contrary, although G_l is also robust against malicious link attacks, its topologies are almost random. Lastly, G_m is rather attractive because it is more balanced. In practice, many special requirements from the real world are often needed, so each kind of network structure obtained by MOEA-RSF$_{MMA}$ is useful for decision makers.

Algorithm 6.6 MOEA-RSF$_{MMA}$

Input:

G_0 Initial SFN;

Ω': Population size for MA-RSF$_{MA}$;

Ω: Population size;

p_c: Crossover rate:

p_l: Local search probability;

α: Predefined parameter for the local search operator;

τ: Sampling interval;

Output:

A set of networks on the Pareto front obtained;

$\mathbf{P}^1 \leftarrow \mathbf{R}_n^D\text{-}\mathbf{Sampling\_phase}(G_0, \Omega', \Omega, p_c, p_l, \alpha, \tau)$;

————————R_n^D-R_l^{BC}-optimization phase————————

Calculate the value of R_l^{BC} for each chromosome in $\mathbf{P}^1$, $t \leftarrow 1$;

while termination criteria are not satisfied **do**

$\mathbf{P}_c^t \leftarrow \varnothing$ /*$\mathbf{P}_c^t$ *is the child population of* $\mathbf{P}^t$*/

repeat

Randomly choose two chromosomes G_i^t and G_j^t that have not been selected;

$(G_{ci}^t, G_{cj}^t) \leftarrow \mathbf{Crossover\text{-}Operator}(G_i^t, G_j^t, p_c)$;

$\mathbf{P}_c^t \leftarrow \mathbf{P}_c^t \bigcup \left\{ G_{ci}^t, G_{cj}^t \right\}$;

until all chromosomes in $\mathbf{P}^t$ have been selected

Calculate the values of R_n^D and R_l^{BC} for each chromosome in $\mathbf{P}_c^t$;

$\mathbf{R}^t \leftarrow \mathbf{P}^t \bigcup \mathbf{P}_c^t$;

$\mathbf{F} \leftarrow \mathbf{Fast\text{-}Non\text{-}dominated\text{-}Sort}(\mathbf{R}^t)$;

$\mathbf{P}^{t+1} \leftarrow \varnothing$;

$i \leftarrow 0$;

while $|\mathbf{P}^{t+1} \bigcup \mathbf{F}_i| \leq \Omega$ **do**

$\mathbf{P}^{t+1} \leftarrow \mathbf{P}^{t+1} \bigcup \mathbf{F}_i$;

$i \leftarrow i + 1$;

end while

$\mathbf{Crowding\text{-}Distance\text{-}Assignment}(\mathbf{F}_i)$;

Sort $\mathbf{F}_i$ in decreasing order according to the crowding distance;

$\mathbf{P}^{t+1} \leftarrow \mathbf{P}^{t+1} \bigcup \mathbf{F}_i \left[1 : \left(\Omega - |\mathbf{P}^{t+1}| \right) \right]$;

$t \leftarrow t + 1$;

end while

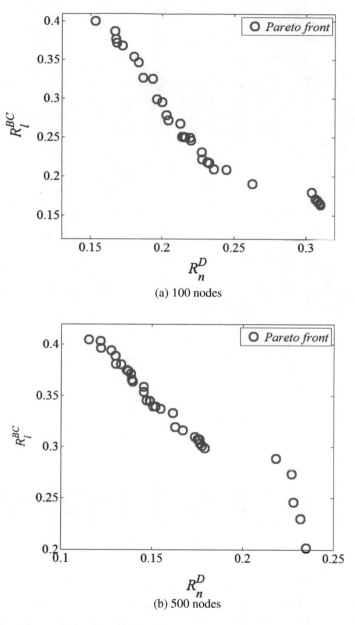

(a) 100 nodes

(b) 500 nodes

Fig. 6.6 Pareto fronts obtained by MOEA-RSF$_{MMA}$ for synthetic SF networks with size of (**a**) 100 nodes and (**b**) 500 nodes

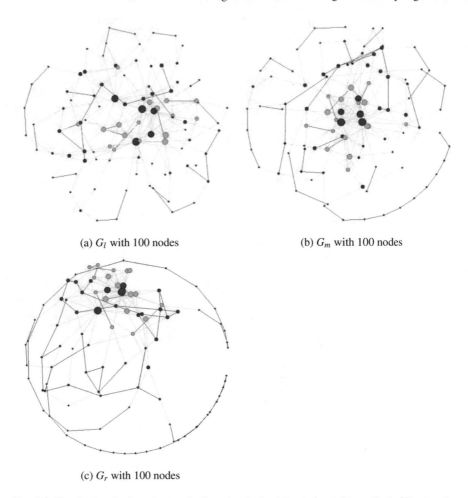

(a) G_l with 100 nodes   (b) G_m with 100 nodes

(c) G_r with 100 nodes

Fig. 6.7 Topologies of selected networks from the obtained Pareto fronts for synthetic SF networks with 100 nodes. The size of node is proportional to the node degree, and edges connecting nodes with the same degree are highlighted

References

1. Adamic, L.A., Huberman, B.A.: Power-law distribution of the world wide web. Science **287**(5461), 2115–2115 (2000)
2. Albert, R., Barabási, A.L.: Statistical mechanics of complex networks. Rev. Mod. Phys. **74**(1), 47 (2002)
3. Albert, R., Jeong, H., Barabási, A.L.: Error and attack tolerance of complex networks. Nature **406**, 378–382 (2000)
4. Barabási, A.L., Albert, R.: Emergence of scaling in random networks (1999)
5. Bauer, D., Boesch, F., Suffel, C., Tindell, R.: Connectivity extremal problems and the design of reliable probabilistic networks. In: The Theory and Application of Graphs, pp. 89–98 (1981)

6. Boesch, F., Frisch, I.: On the smallest disconnecting set in a graph. IEEE Trans. Circ. Theor. **15**(3), 286–288 (1968)
7. Bonacich, P.: Some unique properties of eigenvector centrality. Soc. Netw. **29**(4), 555–564 (2007)
8. Borgatti, S.P.: Centrality and network flow. Soc. Netw. **27**(1), 55–71 (2005)
9. Brandes, U.: On variants of shortest-path betweenness centrality and their generic computation. Soc. Netw. **30**(2), 136–145 (2008)
10. Buesser, P., Daolio, F., Tomassini, M.: Optimizing the robustness of scale-free networks with simulated annealing, 167–176 (2011)
11. Cohen, R., Erez, K., Ben-Avraham, D., Havlin, S.: Resilience of the internet to random breakdowns. Phys. Rev. Lett. **85**(21), 4626 (2000)
12. Cohen, R., Erez, K., Ben-Avraham, D., Havlin, S.: Breakdown of the internet under intentional attack. Phys. Rev. Lett. **86**(16), 3682 (2001)
13. Deb, K., Pratap, A., Agarwal, S., Meyarivan, T.: A fast and elitist multiobjective genetic algorithm: NSGA-II. IEEE Trans. Evol. Comput. **6**(2), 182–197 (2002)
14. Dorogovtsev, S.N., Mendes, J.F.: Evolution of networks: from biological nets to the internet and WWW (2013)
15. Estrada, E.: Network robustness to targeted attacks. The interplay of expansibility and degree distribution. Eur. Phys. J. B Condens. Matter Complex Syst. **52**(4), 563–574 (2006)
16. Estrada, E., Higham, D.J., Hatano, N.: Communicability betweenness in complex networks. Phys. A Stat. Mech. Appl. **388**(5), 764–774 (2009)
17. Fiedler, M.: Algebraic connectivity of graphs. Czech. Math. J. **23**(2), 298–305 (1973)
18. Frank, H., Frisch, I.: Analysis and design of survivable networks. IEEE Trans. Commun. Technol. **18**(5), 501–519 (1970)
19. Hage, P., Harary, F.: Eccentricity and centrality in networks. Soc. Netw. **17**(1), 57–63 (1995)
20. Harary, F.: Conditional connectivity. Networks **13**(3), 347–357 (1983)
21. Herrmann, H.J., Schneider, C.M., Moreira, A.A., Andrade Jr, J.S., Havlin, S.: Onion-like network topology enhances robustness against malicious attacks. J. Stat. Mech. Theory Exp. **2011**(01) (2011). (DOI P01027)
22. Krasnogor, N., Smith, J.: A tutorial for competent memetic algorithms: model, taxonomy, and design issues. IEEE Trans. Evol. Comput. **9**(5), 474–488 (2005)
23. Latora, V., Marchiori, M.: Efficient behavior of small-world networks. Phys. Rev. Lett. **87**(19) (2001). (DOI 198701)
24. Louzada, V.H., Daolio, F., Herrmann, H.J., Tomassini, M.: Smart rewiring for network robustness. J. Complex Netw. **1**(2), 150–159 (2013)
25. Louzada, V.H., Daolio, F., Herrmann, H.J., Tomassini, M.: Generating robust and efficient networks under targeted attacks. In: Propagation Phenomena in Real World Networks, pp. 215–224. Springer (2015)
26. Moscato, P., et al.: On evolution, search, optimization, genetic algorithms and martial arts: towards memetic algorithms. Caltech concurrent computation program, C3P Report, vol. 826 (1989)
27. Newman, M.E.: Mixing patterns in networks. Phys. Rev. E **67**(2) (2003)
28. Ong, Y.S., Keane, A.J.: Meta-lamarckian learning in memetic algorithms. IEEE Trans. Evol. Comput. **8**(2), 99–110 (2004)
29. Paul, G., Sreenivasan, S., Stanley, H.E.: Resilience of complex networks to random breakdown. Phys. Rev. E **72**(5) (2005). (DOI 056130)
30. Dawkins, R.: The Selfish Gene. Oxford University Press (1976)
31. Sampaio Filho, C.I., Moreira, A.A., Andrade, R.F., Herrmann, H.J., Andrade Jr, J.S.: Mandala networks: ultra-small-world and highly sparse graphs. Scientific reports, vol. 5 (2015)
32. Schneider, C.M., Moreira, A.A., Andrade, J.S., Havlin, S., Herrmann, H.J.: Mitigation of malicious attacks on networks. In: Proceedings of the National Academy of Sciences, vol. 108, pp. 3838–3841. National Acad Sciences (2011)
33. Watts, D.J., Strogatz, S.H.: Collective dynamics of 'small-world' networks. Nature **393**(6684), 440 (1998)

34. Wu, J., Barahona, M., Tan, Y.J., Deng, H.Z.: Spectral measure of structural robustness in complex networks. IEEE Trans. Syst. Man Cybern. Part A Syst. Hum. **41**(6), 1244–1252 (2011)
35. Xiao, S., Xiao, G., Cheng, T., Ma, S., Fu, X., Soh, H.: Robustness of scale-free networks under rewiring operations. EPL (Europhys. Lett.) **89**(3) (2010). (DOI 38002)
36. Zeng, A., Liu, W.: Enhancing network robustness against malicious attacks. Phys. Rev. E **85**(6) (2012). (DOI 066130)
37. Zhou, M., Liu, J.: A memetic algorithm for enhancing the robustness of scale-free networks against malicious attacks. Phys. A Stat. Mech. Appl. **410**, 131–143 (2014)
38. Zhou, M., Liu, J.: A two-phase multiobjective evolutionary algorithm for enhancing the robustness of scale-free networks against multiple malicious attacks. IEEE Trans. Cybern. **47**(2), 539–552 (2017)

Chapter 7
Other Optimization Problems
in Complex Networks

There are many optimization problems need to be solved in the field of complex networks. In previous chapters, we have introduced how to use EAs to solve community detection problems and network robustness optimization problems. In this chapter, we introduce two other optimization problems related to complex networks, namely improving cooperation level of evolutionary games (EGs) on networks and network reconstruction.

7.1 Improving Cooperation Level of Evolutionary Games on Networks

Game theory is a discipline that studies the competitive relationship among individuals attracts more and more attention in recent decades [5, 6, 8, 15, 17–19, 25]. A very important part of the game theory research is to study the emergence and persistence of cooperation behavior since cooperation is essential to the evolutionary process of species. Among so many models used in game theory, the prisoner's dilemma game (PDG) is one of the most widely investigated games simulating biological evolution in nature and explaining the cooperation behavior among selfish individuals. Nowak et al. first employed the spatial structure into the PDG, and individuals are constrained to play only [18] with their immediate neighbors. Their simulation results showed that topology constraints influence the evolution of cooperation and this has also been confirmed years later.

Following the breakthrough research done by Nowak et al. combining the spatial structure with the game theory and the rapid development of the research on complex networks, many further researches have been done. Santos et al. in [20] found that graph topology plays a determinate role in the evolution of cooperation. Hauert et al. designed simulations to study the PDG and the Snowdrift Game (SDG) in lattice networks [13], from the simulation results of which we can see that the cooperators in the PDG are likely to form clusters to defend against defectors. Santos et al. have studied the PDG on scale-free networks and found that the heterogeneity of networks

© Springer Nature Switzerland AG 2019
J. Liu et al., *Evolutionary Computation and Complex Networks*,
https://doi.org/10.1007/978-3-319-60000-0_7

favors the emergence of cooperation [21]. However, further research on the PDG in community networks showed that heterogeneity of networks does not always improve cooperation frequency [5]. The investigation on the cooperation level of Holme and Kim networks (HKNs) showed a positive correlation between clustering coefficient and cooperation frequency [1].

Most researches described above only studied the cooperation level of special designed network models without providing a general conclusion in optimizing the cooperation level of a given network. As various factors found or not found yet in the PDG can influence the corresponding cooperation level, it is hard for a special designed network model to take so many effected factors into account. Although cooperation frequency is influenced by many factors, there we considered should always be an ideal topology of network that is of higher cooperation frequency when other effected factors are given. Moreover, researchers have proposed different possible mechanisms to explain cooperation in nature. Kin selection indicates that helping group members eventually helps the individual itself. And group selection is proposed to illustrate that cooperative groups may be more likely to survive than a less cooperative one. Thus, a method to improve cooperation level of network is helpful to further investigate the cooperation behavior in nature. The optimization algorithms that can adjust the topology of a given network to promote the cooperation level are in hard demand. With the networks promoted, we can further investigate the characteristic of highly cooperative networks in a certain gaming environment. Moreover, since cooperative groups are found out be more likely to survive in group selections of nature than a less cooperative one, such optimization algorithms may help us to produce a group with tenacious vitality.

7.2 Network Reconstruction

One of the outstanding problems in interdisciplinary science is how to identify, predict, and control nonlinear and complex systems. Much evidence has shown that interaction patterns among dynamics elements captured by complex networks play an important role in controlling the collective dynamics [23]. However, a great challenge is that the network structure and the nodal dynamics are often unknown, and instead, only the limited observed time series are available. Reconstructing complex network structure and dynamics from measureable data has become a central issue in contemporary network science and engineering [11, 12]. Typical examples include evolutionary games networks [11, 12, 24], propagation networks [22], gene regulatory networks [4–10].

An important class of collective dynamics is evolutionary games in the human society. For criminal gangs, the police need to master the relationships between the members, namely agent-to-agent networks. However, in the real life, it is difficult to directly access to this network, and maybe only the payoff and strategy of its members are available. Therefore, how to reconstruct the agent-to-agent networks from these available information, namely profit sequences, is worth being studied.

Recent efforts have focused on the inverse problem of EG networks where the network reconstruction problem (NRP) is converted into a sparse signal reconstruction problem that can be solved by exploiting sparse learning algorithms, such as the lasso and compressed sensing [11, 12, 24]. In particular, reconstructing the whole network structure can be achieved by inferring local connections of each node individually. The problem of local structure reconstruction incorporates both the natural sparsity of complex networks and measurement error (the difference between observed data and simulated data). This problem is typically solved by using sparse learning algorithms which transform two objectives into one objective by multiplying each objective with a weighting factor and then summing up all contributions. The choice of the weighting factor has a great impact on the performance of sparse learning methods. However, a shortcoming of these penalty approaches is that it is not easy to determine this key parameter which can maximize the performance. Moreover, it is impossible to conduct the cross-validation to obtain the optimal values of this key parameter, especially when given limited data disturbed by noise and unexpected factors is not enough to use split testing data from them. Sometimes, there is also no gold standard to implement the cross-validation. Last but not least, playing cross-validation for the lasso is time-consuming for large-scale problems. Thus, a robust and completely data-driven approach for solving this problem has yet to be discovered.

Focusing on the problem, a multi-objective network reconstruction (MNR) framework to cope with the network reconstruction problem from profit sequences based on multi-objective evolutionary algorithm (MOEA), termed as MOEANet, was proposed in [26]. To overcome the shortcoming of penalty approaches, the problem of local structure reconstruction is first modeled as a multi-objective optimization problem (MOP). One objective is to minimize the difference between the input data and the simulated data; the other is to search for sparse structure. All solutions in the Pareto set are optima of MOPs and represent different levels of compromise between the competing objectives. Thus, these solutions with different properties can be provided to decision makers. However, sometimes, it is necessary to determine which solution in a Pareto set (PS) is the best. Knee regions [2, 7], where further improvement in one objective causes a rapid degradation in other objectives, have attracted considerable interest in the study of MOPs and decision makers have been shown to prefer solutions that lie in knee regions. Therefore, an angle-based method [2, 16] is employed to select the eclectic Pareto solution from the Pareto front (PF) produced by EAs. Finally, the whole network can then be assembled by simply matching neighboring sets of all nodes.

7.2.1 Network Reconstruction from Profit Sequences

During the evolution of EG, we assume that only the profit sequences of all agents and their strategies at each round are available. In the EG network reconstruction problem (EGNRP), agent-to-agent interactions are learned from profit sequences.

The key to solve the EGNRP lies in the relationship between the agents' payoffs and strategies. The interactions among agents in the network can be characterized by a $N \times N$ adjacency matrix X with elements $x_{ij} = 1$ if agents i and j are connected, and $x_{ij} = 0$ otherwise. Also, the interactions can be generalized straightforwardly to the weighted networks. Using the weights to characterize various interaction strengths, we define the weighted adjacency matrix X as: If i connects to j, $x_{ij} \geq 1$; otherwise, $x_{ij} = 0$. The payoff of agent i can be expressed by

$$Y_i(t) = \sum_{l=1}^{N} x_{il} S_i^T(t) P S_l(t) \tag{7.1}$$

where x_{il} ($l = 1, 2, ..., N$) represents a possible connection between agent i and its neighbor l; $X_{il} S_i^T(t) P S_l(t)$ ($l = 1, 2, ..., N$) stands for the possible payoff of agent i from the game with agent i; and $t = 1, 2, ..., m$ is the number of rounds that all agents play the game with their neighbors. The relationship among the vector Y_i, the matrix A_i, and the neighbor-connection vector X_i of agent i is described as follows:

$$Y_i = A_i \times X_i \tag{7.2}$$

where

$$Y_i = (Y_i(1), Y_i(2), ..., Y_i(m))^T \tag{7.3}$$

$$X_i = (x_{i1}, ..., x_{i,i-1}, x_{i,i+1}..., x_{iN})^T \tag{7.4}$$

$$A_i = \begin{pmatrix} D_{i1}(1) & ... & D_{i,i-1}(1) & D_{i,i+1}(1) & ... & D_{iN}(1) \\ D_{i1}(2) & ... & D_{i,i-1}(2) & D_{i,i+1}(2) & ... & D_{iN}(2) \\ \vdots & \ddots & \vdots & & \vdots & \ddots & \vdots \\ D_{i1}(m) & ... & D_{i,i-1}(m) & D_{i,i+1}(m) & ... & D_{iN}(m) \end{pmatrix} \tag{7.5}$$

where $D_{x,y}(t) = S_x^T(t) P S_y(t)$. Y_i can be obtained directly from the payoff data and A_i can be calculated from the strategy data. In a similar fashion, the neighbor-connection vectors of all other agents can be predicted, yielding the network adjacency matrix $X = (X_1, X_2, ..., X_n)$.

Our goal is to reconstruct X_i from Y_i and A_i. Thus, the measurement error needs to be minimized. Note that the number of nonzero elements in X_i, i.e., on average the number of real connections of node i, is much less than the number of all possible connections. This indicates that X_i is sparse, which is ensured by the natural sparsity of complex networks. There are many methods to solve this EGNRP by handling the following problem [11, 12, 24]:

$$\min_{X_i} \left(\frac{1}{2m} \parallel A_i X_i - Y_i \parallel_2^2 + \lambda \parallel X_i \parallel_1 \right) \tag{7.6}$$

where λ is a constant that controls the trade-off between the measurement error and the sparsity of networks. The L_1 norm ensures the sparsity of structure; simultaneously, error control term ensures the robustness of NR against noise.

7.2.2 MNR Model in Evolutionary Games

To balance the importance of measurement error with respect to the sparsity of networks, a trade-off parameter has to be determined. A shortcoming of this method is that it introduces a parameter λ, and with different values of λ, different optimal results can be achieved. The constant is usually determined by trial and error. It is time-consuming to use the method of trial and error owing to the sizes of both the network, and the dataset is huge. Moreover, because of the absence of gold standards of real-world network structure, the lasso cannot use the cross-validation to obtain the optimal value of λ. Furthermore, when there are no enough data, we cannot use split testing data from raw data, especially when the performance of the method is seriously affected by the amount of data. One way of avoiding the choice of λ is to convert the problem into MOPs [3–14]. By analyzing the relationship between the Pareto optimal vectors distributed on the PF, an appropriate solution is selected from the Pareto optimal set. Considering the measurement error and the sparsity of network as two objectives, we establish the MNR model as follows:

$$\min_{X_i} \left(\frac{1}{m} \parallel A_i X_i - Y_i \parallel_2^2, \parallel X_i \parallel_1 \right) \tag{7.7}$$

Then, a multi-objective EA, termed as MOEANet, was proposed to solve the above MNRP in [26]. MOEANet can return a set of solutions on the PF. Each point on the PF represents a certain local network structure. To find the best solution to decision makers, a selection strategy based on knee regions was employed. Knee regions are solutions that have the maximum marginal rates of return, i.e., for which an improvement in one objective causes a severe degradation in another. An angle-based method [2, 16], for locating the knee regions on the PF, is considered.

References

1. Assenza, S., Gómez-Gardeñes, J., Latora, V.: Enhancement of cooperation in highly clustered scale-free networks. Phys. Rev. E **78**(1) (2008). DOI 017101
2. Branke, J., Deb, K., Dierolf, H., Osswald, M., et al.: Finding knees in multi-objective optimization **3242**, 722–731 (2004)
3. Cai, Z., Wang, Y.: A multiobjective optimization-based evolutionary algorithm for constrained optimization. IEEE Trans. Evol. Comput. **10**(6), 658–675 (2006)
4. Chang, Y.H., Gray, J.W., Tomlin, C.J.: Exact reconstruction of gene regulatory networks using compressive sensing. BMC Bioinf. **15**(1), 400 (2014)

5. Chen, X., Fu, F., Wang, L.: Prisoner's dilemma on community networks. Phys. A Stat. Mech. Appl. **378**(2), 512–518 (2007)
6. Chiong, R., Kirley, M.: Effects of iterated interactions in multiplayer spatial evolutionary games. IEEE Trans. Evol. Comput. **16**(4), 537–555 (2012)
7. Deb, K., Gupta, S.: Understanding knee points in bicriteria problems and their implications as preferred solution principles. Eng. Optim. **43**(11), 1175–1204 (2011)
8. Fehr, E., Fischbacher, U.: The nature of human altruism. Nature **425**(6960), 785 (2003)
9. Feizi, S., Marbach, D., Médard, M., Kellis, M.: Network deconvolution as a general method to distinguish direct dependencies in networks. Nature Biotechnol. **31**(8), 726 (2013)
10. Gardner, T.S., Di Bernardo, D., Lorenz, D., Collins, J.J.: Inferring genetic networks and identifying compound mode of action via expression profiling. Science **301**(5629), 102–105 (2003)
11. Han, X., Shen, Z., Wang, W.X., Di, Z.: Robust reconstruction of complex networks from sparse data. Phys. Rev. Lett. **114**(2) (2015). DOI 028701
12. Han, X., Shen, Z., Wang, W.X., Lai, Y.C., Grebogi, C.: Reconstructing direct and indirect interactions in networked public goods game. Sci. Rep. **6** (2016). DOI 30241
13. Hauert, C., Doebeli, M.: Spatial structure often inhibits the evolution of cooperation in the snowdrift game. Nature **428**(6983), 643 (2004)
14. Kan, W., Jihong, S.: The convergence basis of particle swarm optimization. In: 2012 International Conference on Industrial Control and Electronics Engineering (ICICEE), pp. 63–66. IEEE (2012)
15. Li, J., Kendall, G.: The effect of memory size on the evolutionary stability of strategies in iterated prisoner's dilemma. IEEE Trans. Evol. Comput. **18**(6), 819–826 (2014)
16. Li, L., Yao, X., Stolkin, R., Gong, M., He, S.: An evolutionary multiobjective approach to sparse reconstruction. IEEE Trans. Evol. Comput. **18**(6), 827–845 (2014)
17. Nowak, M.A.: Five rules for the evolution of cooperation. Science **314**(5805), 1560–1563 (2006)
18. Nowak, M.A., May, R.M.: Evolutionary games and spatial chaos. Nature **359**(6398), 826–829 (1992)
19. Nowak, M.A., Sigmund, K.: Evolution of indirect reciprocity. Nature **437**(7063), 1291–1298 (2005)
20. Santos, F., Rodrigues, J., Pacheco, J.: Graph topology plays a determinant role in the evolution of cooperation. Proc. R Soc. London B Biol. Sci. **273**(1582), 51–55 (2006)
21. Santos, F.C., Pacheco, J.M.: Scale-free networks provide a unifying framework for the emergence of cooperation. Phys. Rev. Lett. **95**(9) (2005). DOI 098104
22. Shen, Z., Wang, W.X., Fan, Y., Di, Z., Lai, Y.C.: Reconstructing propagation networks with natural diversity and identifying hidden sources. Nat. Commun. **5** (2014)
23. Strogatz, S.H.: Exploring complex networks. Nature **410**(6825), 268 (2001)
24. Wang, W.X., Lai, Y.C., Grebogi, C., Ye, J.: Network reconstruction based on evolutionary-game data via compressive sensing. Phys. Rev. X **1**(2) (2011). DOI 021021
25. Wedekind, C., Milinski, M.: Cooperation through image scoring in humans. Science **288**(5467), 850–852 (2000)
26. Wu, K., Liu, J., Wang, S.: Reconstructing networks from profit sequences in evolutionary games via a multiobjective optimization approach with lasso initialization. Sci. Rep. **6** (2016). DOI 37771

Index

© Springer Nature Switzerland AG 2019
J. Liu et al., *Evolutionary Computation and Complex Networks*,
https://doi.org/10.1007/978-3-319-60000-0

Printed in the United States
By Bookmasters